T0251703

Practical Sanitation
in the Food Industry

Practical Sanitation
in the
Food Industry

IAN S. MADDOX

Department of Process and Environmental Technology
Massey University, Palmerston North
New Zealand

 Gordon and Breach Science Publishers

Switzerland • Australia • Belgium • France • Germany • Great Britain
India • Japan • Malaysia • Netherlands • Russia • Singapore • USA

Copyright © 1994 by OPA (Amsterdam) B.V.

All rights reserved.

No part of this book may be reproduced or utilized in any form or by any means, electronic or mechanical, including photocopying and recording, or by any information storage or retrieval system, without permission in writing from the publisher. Printed in Singapore by Stamford Press Pte Ltd.

Gordon and Breach Science Publishers S.A.
Avenue Gratta-Paille 2
Case postale 531
1000 Lausanne 30 Grey, Switzerland

British Library Cataloguing in Publication Data

Maddox, Ian S.
  Practical Sanitation in the Food Industry
  I. Title
  363.1927024642

  ISBN 2-88124-992-2 (softcover)
  ISBN 2-88449-005-1 (hardcover)

CONTENTS

PREFACE

The food industry is one of the largest in the world, and on a global basis, employs several million people. The performance of this industry is, literally, vital to everyone's health.

Over recent years, the industry has undergone a major transformation. The concept and practice of food processing has expanded so that less preparation is conducted in the home, and food products are increasingly transported around the world to be consumed out of season.

Unfortunately, food can be a vehicle for the transmission of disease, and, in addition, spoilage may occur if products are transported long distances or held in storage for any length of time. Hence, the industry must produce, store and transport its products in such a way that they remain attractive and safe to the consumer. Personnel employed in the food industry therefore need to have a sound education and training in food hygiene and sanitation. For this reason, many textbooks have been written, particularly at university level, to ensure that our microbiologists and managers are fully informed about disease-causing microorganisms and safety management programs. However, there is a serious shortage of books aimed at general supervisory staff in the industry, or at non-microbiologists. In particular, books describing the practical aspects of developing, implementing, evaluating and managing a sanitation program seem to be lacking. It is the purpose of this book to fill this gap, and to present a text which less academically-inclined personnel will find useful. At the same time, the book is intended for tertiary students who are not majoring in microbiology, but who intend to work in the food industry, e.g. chemists or engineers. I have tried to keep detailed technical aspects to a minimum, but there may be occasions where I have failed to do this. This will not, however, detract from the general message of the text.

For the preparation of this book, I thank the many students who have provided feedback to my lecture courses, and the many industry personnel who have commented on my approach. I also thank Maureen Oemcke for word processing, Patty Ratumaitavuki for preparing the diagrams and my wife Noemi for providing encouragement throughout.

Ian S. Maddox

Chapter 1

INTRODUCTION

This book is concerned with the principles and practice of sanitation in the food industry. Sanitation comprises those procedures which are employed to achieve the production of a hygienic food product. Hygienically-produced food will be:

- wholesome   (free from disease and disease-producing organisms)
- fresh   (free from spoilage and large numbers of potential spoilage organisms)
- clean   (free from visible dirt).

    Hence, the purpose of sanitation is to prevent contamination of the product (or raw material or ingredient) with

- visible dirt   (this can lead to immediate rejection of the product)
- spoilage organisms   (these can cause premature spoilage of the product and thus shorten shelf-life)
- pathogenic organisms (these can cause disease in the consumer, and thus represent a health hazard).

Sanitation is an integral part of **quality assurance**. This can be defined as the precautions which are taken to ensure that the product will pass quality control examination. Because most methods of quality control examination involve destructive testing (i.e. the sample which is tested is destroyed, and does not reach the consumer), the examination really assesses the **process**. Hence, the old saying "... if the process is clean, the product will be clean"; i.e. if food is processed in a good, hygienic manner, then we can have confidence in the quality of the product that reaches the consumer.

Contamination of a food can be with either visible dirt or invisible dirt. The latter type is due to microorganisms. In reality, control of microorganisms during food processing involves two aspects:

1. Prevention of contamination by the adoption of sound hygienic practices, i.e. good sanitation procedures.

2. Prevention of proliferation of microorganisms on the food by employing conditions which restrict their growth.

Although the goal of sanitation is to completely prevent contamination, nevertheless it often occurs. In this case, subsequent prevention of proliferation, or even killing of the microorganisms, becomes important. While this is not the main subject of this book, it is discussed briefly in Chapter 4. Unfortunately, it is often easy to become lax about sanitation in the belief that subsequent treatment, e.g. freezing or heating of the product, will prevent any permanent damage. But this is not true! Even a very small number of microorganisms on a product can grow and cause considerable damage before the product is frozen or heated. Microbial damage, once done can never be repaired! Furthermore, treatments such as freezing cannot be guaranteed to have a killing effect on a microbial population, and once the product is thawed, microbial growth continues.

Sanitation, therefore, is the prevention of contamination of the raw material, ingredient or product. To be effective, we must identify the potential **sources** of contamination so that we can target our efforts accurately. These sources include:

- people
- insects, rodents
- contaminated raw material/ingredients
- air, dust
- dirty equipment and surroundings
- water.

Sanitation procedures are designed to minimise the extent of contamination arising from each of these sources.

Sometimes, sanitation is seen as a cost which does not provide any immediate financial return to a company. This is a very short-sighted view if a company wishes to achieve a long-term reputation as a producer of high-quality product. Good sanitation procedures are a long-term investment which result in:

- extended storage life of product
- attainment of processing conditions which allow inspections by appropriate agencies without any risk of suspension on sanitary grounds, and which instil confidence in the customer
- reduction in risk of product being involved in outbreaks of food poisoning
- fewer product rejections, returns or complaints
- less need to reprocess product.

In an increasingly competitive world, a reputation for quality is of paramount importance.

Occasionally, the question is asked why so much attention is paid to hygiene and sanitation in the modern food industry. It is argued that few special precautions were taken in the past, and no harm came of it then. However, this is a fatuous argument. People today are considerably more health conscious than they used to be, and they demand, quite rightly, that the food they eat be safe. In addition, in developed countries, there is an increasing trend for the busy consumer to purchase "convenience" foods which require minimal preparation in the home. Thus, they rely on the food industry, rather than themselves, to process the raw materials to a safe product. Nevertheless, despite the increased attention to sanitation in the food industry, the incidence of food-borne disease is increasing! There is little doubt that individual companies must continue and expand their efforts to implement strong sanitation programmes.

Along with the increasing attention being paid to sanitation has come an increasing supply and complexity of chemicals and equipment to be used in cleaning procedures. It is sometimes difficult for practitioners to decide which chemical to use and which cleaning procedure to adopt. It is hoped that the information given in this book will assist in making these decisions, and help in the organisation of total sanitation programmes. The emphasis throughout is on the practical aspects, but, wherever appropriate, attention is paid to the background theory so that the practitioner can understand the reasons behind a particular practice.

Chapter 2

THE NATURE OF
MICROORGANISMS

Microorganisms, as the name suggests, are small living things, and the study of them is known as microbiology. Many microorganisms are beneficial to mankind, and life as we know it would not exist without their activities. They are responsible in nature for much of the recycling of plant and animal remains, thereby generating nutrients to allow growth of new crops and animals. However, some microorganisms can cause disease, and this is a major concern to the food industry since they can be transmitted *via* food. Microorganisms which can cause disease are known as **pathogens**, and it is imperative that food is processed in such a manner that pathogens are absent from it. Other microorganisms can cause food to spoil and rot. Consumption of food that has spoiled will not necessarily cause disease, but the food looks, tastes and smells disagreeable, so that it is aesthetically unpleasant to eat. Spoilage of food by microorganisms is a natural consequence of their activities in the recycling of plant and animal remains.

Although a major aim of the food industry is to eliminate unwanted microorganisms from its premises and products, there are some products which are manufactured using the beneficial activities of microorganisms! Cheese is prepared by allowing certain **bacteria**, known as lactic acid bacteria, to grow in milk. As they grow, they produce lactic acid thereby decreasing the pH of the milk. As the pH decreases, the protein that is present in the milk precipitates. This protein, combined with some fat which precipitates with it, is then processed to become cheese. Some speciality cheeses, e.g. blue vein, are deliberately inoculated with **moulds** which are allowed to grow producing special flavours and the distinctive colour. Sour milk, a delicacy in many countries, and yoghurt, are similarly made by adding acid-producing bacteria to milk. These products have been made for many centuries, and were developed as ways to preserve the milk.

Other products made using the activities of microorganisms are beer and wine. In this case, the organisms involved are **yeasts**, which convert the sugar present in malted barley or grape juice to ethanol, to produce beer or wine, respectively. The production of wine has been used for many centuries as a means of preserving grape juice so that it can be consumed "out of season". Sometimes, however, the beer or wine spoils, and bacteria known as acetic acid bacteria convert the ethanol to acetic acid. The beverage is now known as vinegar, and, because of its low pH value, it can be used to prevent the spoilage of other food products by "pickling".

Hence, in the dairy and alcoholic beverage industries, growth of microorganisms is encouraged, but always in a controlled manner. However, it must be emphasised that if the **wrong** bacteria or yeast gains access to the process, the product will be spoiled. Hence, sanitation procedures are just as important in these industries as they are in any other.

Microorganisms are divided into several groups:

- bacteria
- fungi, including moulds and yeasts
- protozoa
- algae
- viruses.

The ones that concern us most in the food industry are bacteria and fungi, but some protozoa, algae and viruses are also capable of causing disease.

2.1 NOMENCLATURE

All living organisms, including microorganisms have two names. The first name is called the genus name, and the second the species name. For example, the human race is called *Homo sapiens*. *Homo* is the genus name, and *sapiens* the species name.

The yeast that is used to make beer and wine is in the genus *Saccharomyces*. However, there are two different species, one called *Saccharomyces carlsbergensis* and the other *Saccharomyces cerevisiae*. If we mention the name of a particular microorganism several times in the same chapter, the generic name can be shortened, e.g. *S. cerevisiae*. Notice that these names are always written in italics (or underlined if written in longhand). This is simply an international convention.

There are many thousands of different microorganisms, and it is important that we group them in such a way that we can recognise their similarities and differences. Thus closely-related species are grouped into the same genus, as with *S. cerevisiae* and *S. carlsbergensis*. Closely-related genera are grouped into the same family. Thus, *Escherichia coli* and *Salmonella typhimurium* are both in the family Enterobacteriaceae, signifying that there are certain similarities between these bacteria. The science of grouping living organisms is known as taxonomy, or classification, and is particularly important when we wish to identify an organism and its possible source.

Taxonomy is based on several criteria:

- morphology, i.e. the appearance of the microorganism under the microscope
- cultural characteristics, i.e. the appearance of the microorganism when it is grown in the laboratory. For example, the

shape and colour of the colonies when growing on an agar plate
- biochemistry, i.e. the chemical changes that occur when the organism is growing
- cell wall composition.

2.2   BACTERIA

Bacteria are single-celled microorganisms which are very small and cannot be seen with the naked eye unless many millions are present, e.g. slime on meat. Typical dimensions of a bacterial cell are 1 μm by 5 μm (1000 μm = 1 mm). Some bacteria are useful to us in the food industry, e.g. the starter bacteria, sometimes known as lactic acid bacteria, that are used for making cheese. Many others, however, cause disease; for example, some species of the genus *Streptococcus* can cause sore throats, and many species of the genus *Salmonella* can cause symptoms of food poisoning.

When viewed under the microscope the distinctive **shapes** of bacterial cells can be clearly seen. There are three basic shapes:

- the sphere (coccus)
- the rod (bacillus)
- the spiral.

When bacteria reproduce, they divide into two equal parts by a process known as binary fission. Sometimes, the descendant cells remain clumped together, and these clumps can be useful in the identification of certain bacterial groups (Figure 2.1).

Even when viewed under the microscope, bacteria can be difficult to see! To overcome this problem, the cells can be stained so that they appear a distinctive colour. The most commonly used stain is the Gram stain, and bacteria are referred to as being either Gram-positive or Gram-negative. The cell walls of Gram-positive bacteria take up the Gram stain and appear blue under the microscope. In contrast, the cell walls of Gram-negative

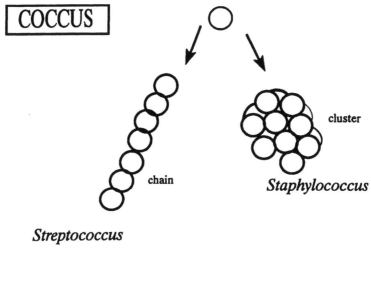

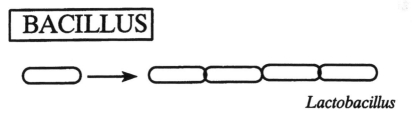

Figure 2.1   Bacterial cells. When cocci divide they can form chains, as in the genus *Streptococcus*, or clusters as in the genus *Staphylococcus*. Bacilli, such as the genus *Lactobacillus*, can form chains of rods. The spiral shaped cells usually remain as individual cells after division.

bacteria do not take up the stain, but are counter-stained to appear pink under the microscope. This Gram reaction reflects the composition of the bacterial cell wall, and is a most important criterion for identification purposes.

The structure of a typical bacterial cell is shown in Figure 2.2. The **cell wall** is responsible for the shape and integrity of the bacterial cell, and is composed of protein, carbohydrate and lipid. Sometimes, outside the cell wall, a slime layer, or capsule, is formed, which causes the slimy feel of spoiled food. Within the cell wall is the **cell membrane**, whose function it is to regulate the uptake of nutrients into the cell, and excretion of end-products. Within the cell membrane is the **cytoplasm**, which constitutes the bulk of the cell. Unlike animal and plant cells, the bacterial cell

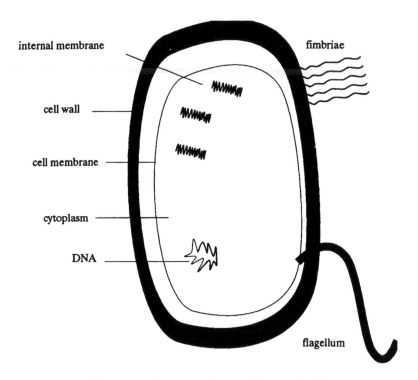

Figure 2.2　Structure of a typical bacterial cell.

does not possess a nucleus, and the DNA (deoxyribonucleic acid) can be found throughout the cytoplasm.

Some bacteria possess one or more **flagella**, which are whip-like structures used in locomotion of the cell. Bacterial cells which possess flagella, and which can rapidly move about, are referred to as being **motile**. Such bacteria can be recognised when growing on agar plates in the laboratory by the spreading of their colonies. The **fimbriae** are very much shorter than flagella, but are much more numerous. They have several functions, including their role in allowing the cell to adhere to a solid surface. They are also used to transfer DNA from one cell to another.

Bacteria in the genera *Bacillus* and *Clostridium* are capable of producing **endospores** (often known simply as spores). These are formed when the cell disintegrates, and they carry all the necessary information and chemicals to be able to germinate and return to the **vegetative cell** form at some future date. Their importance in the food industry is that they are extremely resistant to the procedures that are commonly used to control microorganisms. Thus, they can easily withstand temperatures up to 100°C or higher, and are resistant to many of the chemicals that are used during cleaning procedures. They are particularly important in the canning of foods since if they survive the heating process, they will subsequently germinate and cause the food to spoil and the can to "blow". For this reason, the temperatures and times used during canning are designed to kill spores rather than vegetative cells.

2.3 FUNGI

The fungi consist of **moulds** and **yeasts**. Like bacteria, yeasts are single-celled microorganisms, but although they are considerably larger than bacteria (e.g. 20 µm diameter) they still cannot be seen by the naked eye unless many millions are present. They also differ from bacteria in the way that they reproduce. Bacterial cells divide by simply splitting into two equal halves; yeasts,

Figure 2.3   Reproduction of yeasts by budding.

however, produce daughter cells by a process known as budding (Figure 2.3). A small bud appears on the side of the yeast cell, and slowly grows until it is equal in size to the parent cell. The daughter and parent cells then separate.

Moulds are very different to bacteria. They are not single-celled, but are filamentous microorganisms, i.e. the cells stay attached to each other to form long filaments. When growing on food, or on dirty surfaces, or on agar plates in the laboratory, mould growths are often cream-coloured or colourless. As they grow older, however, many moulds form spores, which may be green, black, blue or some other colour. Hence, mould contamination of food is easily recognised by the coloured spores. Species of *Aspergillus* or *Mucor* often form black spores while *Penicillium* is recognised by blue or green spores.

Moulds and yeasts are only rarely associated with food poisoning symptoms. However, they are common causes of food spoilage, particularly of dry foods such as bread and cheese. Moulds in particular are very easily spread in a factory because of the millions of spores that they produce. It is important, however, to distinguish between these mould spores and the bacterial endospores. Mould spores serve the purpose of rapid spread of the microorganism, but they are not resistant to heat or chemicals. Bacterial endospores, as stated earlier, are very resistant, but do not contribute to rapid spread of the organism.

2.4   PROTOZOA

Protozoa are single-celled microorganisms which differ from bacteria and yeasts in that they do not possess a cell wall. Only a few genera can cause food poisoning symptoms (refer to Chapter 5.3.3), and these are usually transmitted by water, although occasionally food can be involved. Some species, however, have resistant spores, or cysts, which can withstand many cleaning chemicals, and so present a potential problem. There is virtually no involvement of protozoa in food spoilage.

The major hazard to humans from protozoa is from diseases such as malaria. The genus *Plasmodium* is the organism responsible for malaria, and it carries out part of its life cycle in humans, and part in mosquitoes. The mosquito is the vector by which the disease is transmitted from person to person.

2.5 ALGAE

Algae are microorganisms which contain the green pigment chlorophyll. Hence, they are similar to green plants. They vary considerably in size and structure, ranging from single-celled organisms such as *Chlorella* to seaweeds up to 30 metres in length. They play virtually no role in food spoilage or human disease, although some algae do produce toxins which can cause symptoms of food poisoning when ingested (refer to Chapter 5.2.5).

Eutrophication of waterways is a problem which is caused by the presence of excess algae. The algae are stimulated to grow by excessive fertilization of the waterway, usually with phosphate. Subsequently, the algae die and their cells are degraded by bacteria. As they grow, these bacteria consume all the oxygen which is present in the water, causing the death of other aquatic life, including fish. Eventually, the lake can support no plant or animal life, and it is unfit for recreational purposes.

2.6 VIRUSES

Viruses differ from all other microorganisms in that they are not living cells. They are composed only of nucleic acid, e.g. DNA, and protein, and can be considered to be infectious particles which can reproduce only **inside a living cell**. Hence, they are parasites, and they force the host cell to manufacture new viruses which are subsequently liberated to infect new cells. During this process the host cell dies, and so viruses are agents of disease.

There are many types of viruses, some of which infect plants, some infect animals and others even infect microorganisms such as bacteria. The latter type are known as **bacteriophages**, and they are particularly important in the dairy industry where they can infect and destroy the starter bacteria which are used in cheese making. For this reason, dairy companies must practise particularly high standards of hygiene and sanitation to minimise the occurrence of bacteriophages in the factory.

Viruses are responsible for many diseases in humans, e.g. influenza, chickenpox, the common cold and AIDS. Virtually no drugs are available to cure viral diseases, but immunisation is often used to prevent the disease. Unfortunately, vaccines are not available to allow immunisation against all diseases, so once infection has occurred the disease has to run its course.

Viruses are not involved in food spoilage, but many can cause symptoms of food poisoning. Food or water can serve as the vehicle of transmission (refer to Chapter 5.3.2).

Chapter 3

NUTRIENT REQUIREMENTS AND GROWTH OF MICROORGANISMS

All microorganisms require a source of nutrient to enable them to grow. By its very nature, food is an excellent source of nutrient, so we can expect microorganisms to grow on many of our food products unless positive preventative measures are taken. The nutrients required by all organisms can be divided into two groups:

- **macronutrients** are required in concentrations greater than 10^{-4} M, and include carbon, nitrogen, sulfur, phosphorus, magnesium and potassium

- **micronutrients** are required in concentrations less than 10^{-4} M, and include vitamins and trace elements such as zinc, copper, iron, molybdenum, manganese, sodium and calcium.

3.1 MACRONUTRIENTS

3.1.1 Carbon

Microorganisms are classified on the basis of their carbon source:

- **heterotrophs** require organic carbon compounds as their source of carbon, and they obtain their energy by oxidation of these compounds
- **autotrophs** use carbon dioxide as their sole source of carbon, and they obtain their energy requirements from either visible light (photosynthetic autotrophs) or the oxidation of inorganic chemicals (chemosynthetic autotrophs).

In the context of the food industry, we need to be concerned only with heterotrophs, since they are the ones that grow on our food products. Simple sugars such as glucose and sucrose, proteins, lipids and polysaccharides all make excellent carbon sources for heterotrophic microorganisms.

3.1.2   Nitrogen

Nitrogen is required by microorganisms for the synthesis of proteins and nucleic acids. Although some organisms can "fix" atmospheric nitrogen, and many can utilise simple nitrates and ammonium salts, the amino acids and proteins in foods make excellent nitrogen sources for the growth of most microorganisms.

3.1.3   Phosphorus

This is an essential component of nucleic acids, and is a key element in the regulation of metabolism *via* ATP. Most microorganisms can use inorganic phosphates as their source of this nutrient.

3.1.4   Sulfur

This is present in some amino acids, and so is an essential constituent of proteins. Many microorganisms can utilise inorganic sulfate, but others require the sulfur-containing amino acids as a source of this nutrient.

3.1.5   Magnesium

This is a cofactor for some enzymes, and is present in bacterial cell walls. Microorganisms utilise it as the cation, i.e. Mg^{2+}.

3.1.6   Potassium

This is important for the uptake of other metal ions, and is a cofactor for some enzymes. It is utilised as the cation, i.e. K^+.

3.2   MICRONUTRIENTS

These are essential for microbial growth although they are required in only trace amounts. Trace elements such as zinc, copper, iron, molybdenum, manganese, sodium and calcium are readily utilised as their ions. Vitamins can often be synthesized by many microorganisms, but others, known as **fastidious** organisms, need to be supplied with one or more of them. Common vitamins include thiamine (B_1), riboflavin (B_2), pyridoxine (B_6), biotin and folic acid.

3.3   MICROBIOLOGICAL MEDIA

3.3.1   General Purpose Media

One of the major requirements of a microbiology laboratory is for a medium on which most microorganisms of interest will grow. To achieve this, microbiologists have developed media which are undefined in terms of their exact composition, but which contain sufficient concentrations of most macronutrients and micronutrients to allow the growth of a wide range of heterotrophic organisms. These are **general purpose media**, and a common example is nutrient broth which is widely used for the growth of heterotrophic bacteria. Nutrient broth is composed of:

- beef extract; a source of carbohydrate, amino acids, vitamins and salts

- peptone; a source of amino acids, prepared from hydrolysis of proteins
- yeast extract; a source of amino acids, vitamins and salts, prepared from spent brewers' yeast.

To prepare nutrient agar, nutrient broth is solidified using a polysaccharide material known as **agar**.

3.3.2   Special Purpose Media

One of the tasks of a microbiology laboratory is to estimate the numbers of microorganisms present in a food sample. If it is wished to determine the total numbers, general purpose media are used. However, a common requirement is to estimate the numbers of only particular bacterial groups, e.g. Gram negative, enteric bacteria. In this case, a medium is needed which allows the growth of the desired organisms, but not of any others. Hence, **selective**, or **isolation media** are used. An example of such a medium is MacConkey agar, which is used for the enumeration of certain Gram negative, enteric bacteria in food. This medium contains all the nutrients which are required to allow these bacteria to grow, and also contains **selective agents** which inhibit the growth of Gram positive, non-enteric organisms. Crystal violet, at the concentration present in MacConkey agar, inhibits Gram positive bacteria, while bile salts, which occur naturally in the enteric tract of mammals and birds, inhibit non-enteric bacteria.

Another requirement in the microbiology laboratory is for media which can be used to **differentiate** between closely related organisms. For example, on MacConkey agar, it may be wished to enumerate only coliform bacteria, rather than the entire group of Gram negative enterics. By definition, coliforms produce acid when growing on the sugar lactose, while other Gram negative enterics do not. Hence, MacConkey agar contains lactose and an acid-base indicator so that acid-producing and non-acid-producing colonies appear as different colours (usually, red and colourless, respectively). In this way, coliforms can be readily distinguished from their close relatives, and such a medium is known as a **differential** medium.

3.4   GROWTH OF MICROBIAL NUMBERS

In this section, we are concerned with the increase in numbers in a bacterial population which might be present in a food. Bacteria (and yeasts, but not moulds) exist as single cells. When they grow, they increase in size and at the appropriate time divide into two cells (Figure 3.1). These two cells subsequently grow and divide into four cells, and so on. The time taken for a bacterial cell to increase in size and divide into two cells is known as the

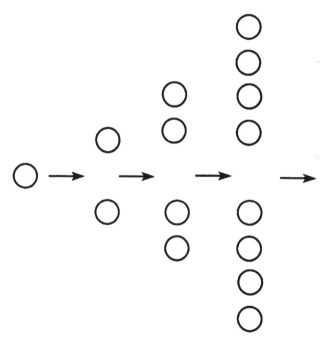

Figure 3.1   Growth of microbial populations. Each cell increases in size, then divides to give two daughter cells. Each daughter cell increases in size and divides to give a total of four cells, and so on.

doubling time. This is defined as the time required to double the number of cells present.

3.4.1   Increase in Cell Numbers

If we assume that a food sample contains one bacterial cell, and that the doubling time is one hour, we can see how the bacterial numbers increase:

Time, hours	Number of cells
0	1
1	2
2	4
3	8
4	16
5	32
10	1 024
24	16 777 216
30	1 073 741 824

Hence, after 30 hours, more than 1 billion bacterial cells are present in the food!

Because these high numbers are laborious to write, microbiologists have adopted a numbering system based on powers of ten. Thus:

$$
\begin{aligned}
1 &= 10^0 \\
10 &= 10^1 \\
100 &= 10^2 \quad (10 \times 10) \\
1000 &= 10^3 \quad (10 \times 10 \times 10) \\
10\,000 &= 10^4 \quad (10 \times 10 \times 10 \times 10) \\[6pt]
10\,000\,000 &= 10^7
\end{aligned}
$$

In the example above, therefore, we would say that after 30 hours of bacterial growth, the number of cells present is approximately 1×10^9.

3.4.2   Effect of the Initial Number of Cells

In the example given above, we started with just one bacterial cell. The following shows the increase in numbers if we start with one, one hundred or one thousand cells, for a doubling time of one hour:

Time, hours	Number of cells		
0	1	100	1 000
4	16	1600	16 000
12	4096	409 600	4 096 000
	(4×10^3)	(4×10^5)	(4×10^6)

This illustrates the importance of minimising the initial numbers of bacteria on the food, since the more bacteria that are present initially, the quicker that the numbers become dangerously high.

3.4.3   Effect of the Doubling Time

The doubling time gives an indication of the suitability of the conditions for bacterial growth. If the conditions can be made less suitable, the doubling time will be increased. The following shows the effect of changing the doubling time from one hour to two hours:

Time, hours		Number of cells	
	Doubling time:	1 hour	2 hours
0		1	1
4		16	4
12		4 096	64
24		16 777 216	4 096
		(1.6×10^7)	(4.1×10^3)

This explains the need in food preservation to select conditions which increase the doubling time and, thus, minimise bacterial growth. This is discussed further in Chapter 4.

3.4.4 Mathematical Expression of Bacterial Growth

The increase in bacterial numbers as a function of time, as described above, can be expressed mathematically as follows:

The increase in bacterial numbers is proportional to the numbers present and the time interval.

i.e. $dN \propto N.dt$ where N is the bacterial number, and t is the time interval.

$dN = \mu.N.dt$ where μ is the specific growth rate, h^{-1}.

$\therefore$ $dN/dt = \mu.N$

$N_t = N_0 e^{\mu t}$

$\ln N = \ln N_0 + \mu t$

If ln N is plotted against t, the slope of the line is equal to the specific growth rate, μ, and the intercept represents the cell numbers at zero time (Figure 3.2). In practice, microbiologists plot $\log_{10}$ rather than ln, so to determine μ the slope must be multiplied by 2.3.

The fact that Figure 3.2 gives a straight line confirms that bacterial growth is exponential, as shown in Section 3.4.1 above.

The specific growth rate, μ, is related to the doubling time, t_d:

$\ln 2 N_0 = \ln N_0 + \mu.t_d$

$\therefore$ $\ln 2 = \mu.t_d$

$t_d = \ln 2/\mu$

$= 0.693/\mu$

3.4.5   Bacterial Numbers on Food

It is not possible to generalise about the acceptable numbers of bacteria on food, since much depends on the specific product involved. However, as a general rule, food will show visible signs of spoilage if the numbers reach 10^9 per gram, or ml, or cm^2 of surface. Pathogenic microorganisms, of course, can cause food poisoning symptoms at concentrations much lower than these.

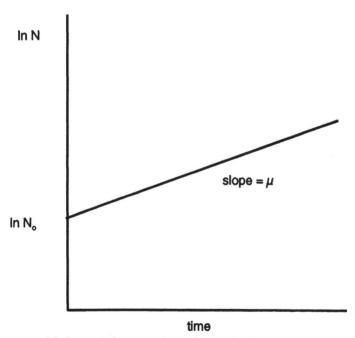

Figure 3.2 Mathematical expression of microbial growth; a plot of the logarithm of the numbers against time. The slope is equal to the specific growth rate, μ.

Chapter 4

EFFECT OF THE PHYSICAL ENVIRONMENT ON MICROORGANISMS

The physical factors discussed in this chapter can affect the growth and/or survival of microorganisms, hence they can be used for microbial control in food or its surroundings. In understanding the affects, it is important to clearly differentiate between "inhibition of growth" (bacteriostatic, fungistatic) and "killing" (bactericidal, fungicidal, germicidal). In the former, growth is inhibited, but the cells are not dead. The cells are dormant, and growth will resume after the constraint has been removed. In contrast, "killing" destroys the cells, and growth can never resume. An object that is completely free of all living microorganisms is said to be **sterile**, and the process is known as sterilisation.

4.1 DEATH KINETICS

Under a specified set of conditions, the number of cells being killed is proportional to the number present and the time interval:

$$dN \propto N.dt$$

where N is the number of cells, and t is the time interval.

$$\therefore dN = -KN.dt$$

where K is a constant, and has a negative sign because the number of cells is decreasing.

$$\frac{dN}{dt} = -KN$$

showing that the kinetics of death are similar to those of a first order chemical reaction, i.e. the rate depends on the concentration of only one reactant, in this case the number of cells present.

$$\ln N = \ln N_0 - Kt$$

where N_0 is the number of cells at zero time, and N is the number surviving after the time interval, t.

If ln N is plotted against time, a straight line results, proving that the killing rate is exponential. In practice, microbiologists usually plot the $\log_{10}$ rather than the natural logarithm (Figure 4.1). Occasionally, the theoretical straight line is a curve showing that some cells in the population are more resistant to the lethal effect than are other cells.

4.2 EFFECT OF TEMPERATURE

As with all chemical reactions, increasing temperature causes an increase in the rate of metabolic processes, and hence the growth rate of microorganisms. Figure 4.2 shows a plot of the growth rate against temperature for any particular microorganism.

The **minimum temperature** is the lowest temperature at which growth occurs. Below this temperature the growth rate is zero.

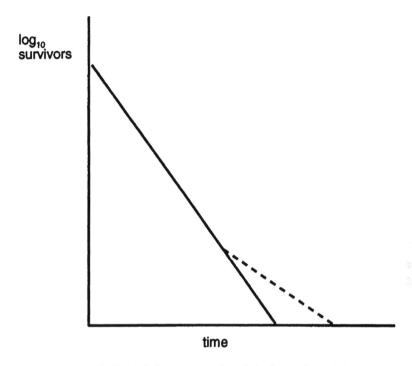

Figure 4.1 Microbial death kinetics. A plot of the logarithm of the surviving cells against time. Occasionally the theoretical straight line is a curve.

Cells become dormant, but do not necessarily die. Growth will resume once the temperature is raised above the minimum temperature.

The **maximum temperature** is the highest temperature at which growth occurs. Above this temperature the growth rate is zero, and vegetative cells will die (although bacterial endospores, which are very resistant to heat, will survive higher temperatures).

The **optimum temperature** is the temperature at which the growth rate is greatest.

Note that the optimum temperature does not lie midway between the minimum and maximum temperature, but is nearer to the maximum. This is due to the effect of temperature on

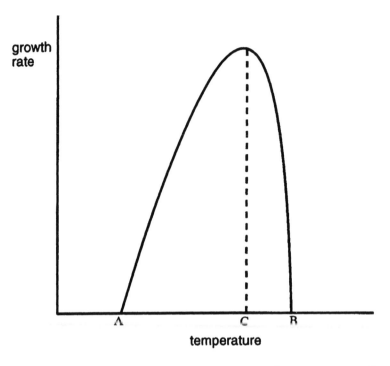

growth rate

temperature

Figure 4.2 Effect of temperature on the growth rate of microorganisms. Point A is the minimum temperature, point B the maximum temperature and point C the optimum temperature.

enzymes. Initially, an increase in temperature leads to increased rates of enzymic reactions, and hence the growth rate. Eventually, however, the temperature gets so high that the enzymes become inactivated. Hence the reactions, and growth, cease. The optimum temperature is a balance between the effect of increasing temperature on reaction rate and enzyme inactivation.

According to their growth temperature relationships, microorganisms are grouped into four categories (Table 4.1). It is important to note that these groups relate to microbial **growth**, and that cells/spores can survive outside these ranges.

In the food industry, thermophiles can sometimes cause problems in canned products, where some cells or spores have

Table 4.1 Groups of microorganisms based on their growth temperature relationships.

Group	Minimum temperature, °C	Optimum temperature, °C	Maximum temperature, °C
Thermophiles	40	55–60	80–90
Mesophiles	10–15	20–40	35–45
Psychrophiles	–5	15	30
Psychrotrophs	–2	30	35–40

survived the heating process. If the cans are not cooled sufficiently quickly, microbial growth and product spoilage may occur.

Most pathogens are mesophiles, so their growth in food can be prevented by refrigeration or freezing. However, a few pathogens, e.g. *Listeria monocytogenes,* are psychrotrophs and can grow at temperatures as low as 3°C.

Psychrophiles and psychrotrophs can cause problems in foods which are preserved by chilling or freezing. It is generally accepted that the lowest temperature at which microbial growth will occur, in practical situations, is –5°C.

4.2.1 Use of High Temperatures to Kill Microorganisms

Several terms are used to quantify the lethal effect of heat on microorganisms.

The **thermal death time** is defined as the time required to kill a culture at a specified temperature. It is a reasonably useful term, but suffers from the major drawback that its value varies depending on the number (or concentration) of cells in the culture. To illustrate this, Figure 4.3 shows the death curves for two different cultures of the same organism. Culture A contained 10^6 cells, and the thermal death time was 6 minutes. Culture B contained 10^4 cells, and the thermal death time was 4 minutes. Hence, it takes a considerably longer time to kill a more concentrated culture. This is an important point in practice when a heat

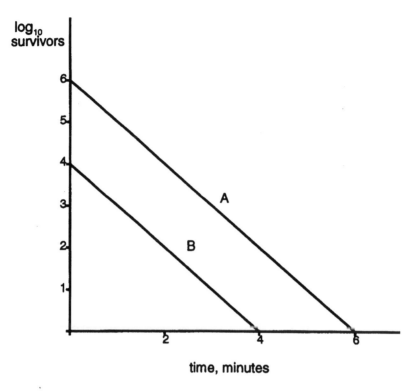

Figure 4.3 Death curves for cultures of different concentrations, showing the effect of concentration on the thermal death time. Culture A, containing 10^6 cells has a thermal death time of 6 minutes, while culture B, containing 10^4 cells, has a thermal death time of 4 minutes.

treatment is used to kill bacteria. Typical examples of thermal death times, for cultures containing 10^7–10^8 cells/ml are given in Table 4.2.

The **decimal reduction time** (D value) is the time required to kill 90% of the cells present in a culture, at a specified temperature. Or, to put it another way, the time required for a one $\log_{10}$ reduction, at a specified temperature (Figure 4.4). Because the death curve is a straight line, this term is independent of the

Table 4.2 Thermal death times of selected organisms, at cell concentrations of 10^7–10^8/ml.

Organism	Temperature °C	Time minutes
Yeast	55	10
Moulds	50	10
Mould spores	60	10
Escherichia coli	60	8
Escherichia coli	80	1
Staphylococcus aureus	65	15
Bacillus subtilis endospores	100	15
Clostridium botulinum endospores	100	300
Clostridium botulinum endospores	120	4

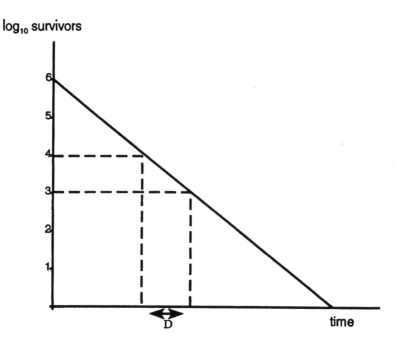

Figure 4.4 Decimal reduction time (D) of a culture.

initial concentration of cells in the culture, and its value will be constant wherever it is measured on the death curve.

If a graph is drawn of the logarithm of the decimal reduction time against the temperature (a thermal resistance curve), the slope of the line gives the **Z value** (Figure 4.5). This is defined as the number of degrees that the temperature must be increased to cause a one $\log_{10}$ reduction in decimal reduction time. A typical value for bacteria would be between 5°C and 10°C.

The decimal reduction time and Z value are used in determining the conditions for thermal processing of foods. The

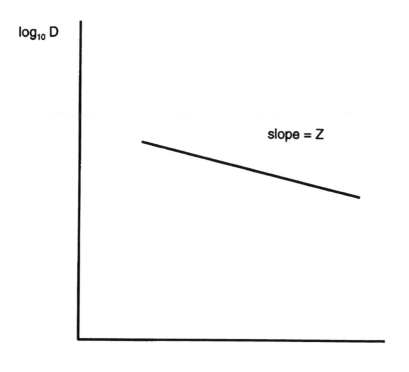

Figure 4.5   Thermal resistance curve of a culture. The slope of the line is the Z value.

decimal reduction time is affected by several factors, which need to be considered when using heat as a lethal agent:

- the higher the temperature, the shorter the time required to kill the cells, i.e. D is lower
- cells are protected by organic materials
- cells are most resistant at neutral pH values, and least resistant under strongly acidic or strongly alkaline conditions
- moist heat is a much more lethal killing agent than is dry heat. Steam is particularly lethal because of its penetrative ability, and its ability to transfer heat by losing its latent heat of vaporisation. For example, to sterilise a moist food, application of a temperature of 120°C for 15 minutes would guarantee sterility. However, to sterilise an anhydrous powder, in the absence of moisture, would require 160°C for 4 hours.

In the food industry, high temperatures are applied in several ways to kill microorganisms. Hot water and steam may be used in cleaning procedures, while one of the purposes of cooking food is to destroy pathogenic microorganisms. However, cooking (or boiling) under normal conditions cannot guarantee the killing of all microorganisms. This is because of the extreme resistance of bacterial endospores. Hence, cooking can be used as a preservative technique only if it is coupled to some other process, e.g. cooking followed by frozen storage at –12°C. In this case, the growth of any cells/spores which survived cooking would be inhibited by the low temperature.

- Canning of food is a process whereby food can be sterilised, and preserved indefinitely. The food is placed in a can, the air is removed by a jet of steam, and the can is sealed. Then, because the can is gas-tight, the contents can be pressurised and heated to temperatures above 100°C, thus destroying bacterial endospores in addition to vegetative cells. After cooling, the canned food can be stored indefinitely.
- Pasteurisation is the application of heat to a food or beverage, usually to temperatures less than 100°C. Pasteurised products

are not necessarily sterile, but the heat treatment has been sufficient to kill certain target organisms, e.g. pathogens. The process is often used for products which would be adversely affected by higher temperatures, e.g. flavour changes in milk, loss of ethanol from beer.

Times and temperatures used in pasteurisation depend on the product, but there are two basic techniques:

- high temperature – short time (HTST) e.g. for milk, 72°C for 15 seconds
- low temperature – hold (LTH) e.g. for beer, 60°C for 20 minutes.

With regard to pasteurisation, care must be taken with **thermoduric organisms**. These are bacteria which, although often mesophiles in terms of their growth, are relatively resistant to pasteurisation temperatures. In the dairy industry, thermoduric organisms include strains of *Lactococcus, Micrococcus* and *Lactobacillus*. Occasionally, they survive pasteurisation and cause premature spoilage of the product.

Because pasteurised products are not necessarily sterile, subsequent storage should be under conditions which minimise or inhibit microbial growth, e.g. pasteurised milk should be stored at refrigeration temperatures.

4.2.2   Use of Low Temperatures to Control Microorganisms

The use of low temperatures is a common preservative technique in the food industry, and it is essential that its effect be understood. For any particular microorganism, if the temperature is decreased to below the minimum temperature, growth will cease, i.e. the cells become dormant. It is possible that some cells will die, but for all practical purposes, chilling and freezing should **never** be considered as techniques for killing microorganisms. (In fact, in microbiology laboratories, freezing is used as a technique for preserving microbial cultures!) To illustrate this point, consider a carcass of meat which has a bacterial load of

$10^7/cm^2$ on its surface, and which is frozen to $-18°C$ for some months. Even if 90% of the bacteria die during storage, the load is still $10^6/cm^2$. Hence, this meat could still spoil rapidly after thawing.

4.2.2.1 Chilling

Chilling is a preservative technique where food is stored at temperatures down to $-1°C$. Because this is above the minimum temperature of many psychrophiles and psychrotrophs, microbial growth does occur in chilled foods. Hence, when used alone, chilling should be considered as only a short-term preservation technique. The actual length of time for which food can be stored at chill temperatures depends on a variety of factors, including:

- the precise temperature of storage
- the nature of the food, e.g. pH, water activity (see later)
- any other preservation technique also being used, e.g. modified atmosphere (see later)
- the initial load of microorganisms.

Chiller management plays an important role in determining the effectiveness of chilling as a preservation technique, where the aim is to restrict microbial growth to a minimum while acknowledging that it does occur. When a chiller is used for cooling the product, rather than for storage, certain guidelines should be followed:

- all chillers must be clean and at the correct temperature before loading with product
- product should be adequately spaced, thus allowing good airflow around surfaces
- chillers should not be partially filled one day, and topped up the next
- chillers must not be loaded beyond their known refrigeration capacity.

Ideally, chilling rates should be as high as possible in order to minimize microbial growth, provided that this is compatible with other aspects of the food quality. For those products that have been cooked, and are then stored at chill temperatures prior to sale and consumption, there is evidence that fast chilling will have a lethal effect on bacteria. However, complete killing of the population must never be assumed.

If there is a breakdown in a chiller's refrigeration or fans, this should always be dealt with urgently, especially if the product temperature is still high. If repairs cannot be completed in a reasonable time, provision should be made to move the food product to an alternative refrigerated area.

4.2.2.2 Freezing

Freezing is a preservation technique where food is stored frozen, normally at temperatures less than –1°C. To prevent all microbial growth, storage should be below –5°C, and preferably as low as –18°C to inhibit adverse chemical reactions.

If frozen food is allowed to warm to temperatures in the range –5°C to –1°C, psychrophilic fungi can sometimes appear. These can be seen as black or green spots, or sometimes as white "whiskers", on the product or container. In fact, growth at these temperatures is very slow, so that if visible growth does appear, it provides evidence that the product temperature has been higher than –5°C for several weeks or months. At temperatures above –2°C, psychrotrophic bacteria, rather than fungi, grow on the product due to their inherently higher growth rates. In this case, the spoilage is usually seen as a slime.

The correct management of freezers and coldstores is an integral part of food quality assurance:

- products should be frozen to the correct temperature before being transferred to the coldstore. There must be adequate airflow over the important cooling surfaces, and transfer to the cold store should take place as quickly as possible under cool, protected conditions;

- coldstores should be clean, and able to hold the food product at the required temperature. Product should not be touching walls, and should be spaced to allow adequate airflow;
- coldstore temperatures must be recorded and monitored regularly. Most coldstores do not maintain the same temperature in all parts, so temperature sensors should be placed in the known "hot spots" (a temperature map should be made of all coldstores). Temperature records are a valuable aid to store management. They detect danger periods when temperatures rise, and provide evidence that food product has not been subjected to undue fluctuations in store temperature.

If fungal growth is seen in a coldstore, it provides clear evidence that temperatures have risen to $-5°C$ for at least several weeks. The cause should be found and the problem rectified. Possible causes include:

- the product was not at the correct frozen temperature before being transferred to the coldstore
- the product was touching a wall, causing a localised heat leak
- temperature fluctuations have occurred, possibly due to doors being left open for long periods.

4.3 pH

As a group, microorganisms can grow over a very wide pH range. However, for any particular microorganism there is usually a well-defined optimum (Figure 4.6). Typical minimum, optimum and maximum pH values for many bacteria and fungi are:

	minimum	optimum	maximum
bacteria	4.5	6.5	8.5
moulds and yeasts	1.5	4.5	7.0

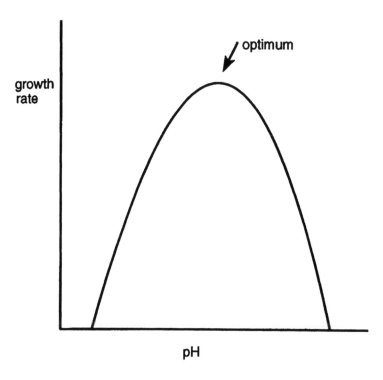

Figure 4.6   Effect of pH on the growth rate of microorganisms.

Thus, moulds and yeasts generally grow under more acidic conditions than do bacteria, and this is reflected in the type of spoilage that occurs on particular foods, e.g. fruit juices, being acidic, are spoiled by yeasts rather than by bacteria.

There are many exceptions to the above figures. For example, the bacterium *Acetobacter* can grow at pH values as low as pH 1. This organism is responsible for the production of acetic acid from ethanol, and is used in the manufacture of vinegar. Other bacteria, known as **alkalophiles** can grow at values up to pH 12, but they are rarely encountered in the food industry.

At pH values below the minimum and above the maximum, microbial cells die rapidly. This is due to the extreme extracellular pH value causing changes to the pH value inside the cells, leading to inactivation of enzymes and subsequent death. An example of this is in the **pickling** of food, where the acetic acid in the vinegar inhibits microbial growth and kills many microbial cells.

4.4   WATER AND WATER ACTIVITY

All microorganisms require water for growth. Vegetative cells contain up to 90% of water, by weight, although spores contain considerably less.

To quantitate the amount of water in a substrate or food which is **available** for microbial growth, we cannot simply measure the amount or concentration of water in the system. Instead, we use the term **Water Activity** (A_w). To understand this concept imagine, first, a sample of pure water. Within this sample, individual water molecules are attracted to each other by weak forces between a hydrogen atom of one water molecule and an oxygen atom of another water molecule. Occasionally, some molecules break free of these forces and escape as water vapour. This causes water to have a measurable vapour pressure. For microorganisms which are present in the water, they must overcome the weak attractive forces to take up water into the cell. Now imagine that a solute, such as salt or sugar, is dissolved in the water sample. Molecules of water are now attracted to the solute molecule by forces greater than those attracting water molecules to each other. It is now more difficult for water molecules to escape as vapour, and so the solution has a lower vapour pressure than does pure water. At the same time, it is more difficult for microorganisms to attract water into the cell, and so the water is considered to be less available.

Water Activity is defined as the ratio of the vapour pressure (P) of the solution or substrate (food) to that of pure water (P_o).

$$A_w = \frac{P}{P_o}$$

Hence, as the solute concentration in the water increases, so the vapour pressure and the water activity decrease. As the water activity decreases, so the availability of water to microorganisms decreases. Table 4.3 shows values of water activity for some salt and sugar solutions, and for some foods.

In general, most bacteria fail to grow at water activity values less than 0.90, yeasts below 0.88, and moulds below 0.80. However, there are exceptions. *Staphylococcus aureus*, a salt-tolerant bacterium which can cause a type of food poisoning, will grow at a value of A_w 0.85. Halophilic bacteria grow in the A_w range 0.75 to 0.91, while osmophilic fungi will grow at values as low as 0.63. Figure 4.7 shows some of these relationships. Note that there is no microbial growth at an A_w value of 1.00, as this is pure water.

It is obvious from the above that manipulation of water activity can be used as a means of inhibiting microbial growth in food. This can be achieved in three ways:

- drying: where water is removed so that the solute concentration increases
- salting: where salt is added to the food to increase the solute concentration, e.g. salted meat, salted fish
- sugar addition: where sugar is used to increase the solute concentration, e.g. jams.

Occasionally, such foods may spoil if they become contaminated with halophilic bacteria or osmophilic fungi. Freezing of

Table 4.3 Water activity values for some salt and sugar solutions, and for some foods.

Pure water	1.00
NaCl (8.7 g/l)	0.99
Sucrose (92 g/l)	0.99
NaCl (162 g/l)	0.90
Meat, fruit, vegetables	0.98
Bread	0.95
Ham	0.90
Cereals	0.70

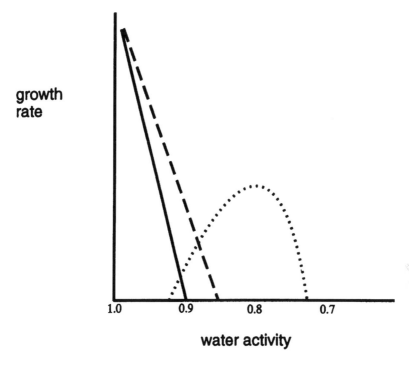

Figure 4.7 Effect of water activity on the growth rate of microorganisms. Most bacteria, ——; salt tolerant bacteria, – – –; halophilic bacteria, ·······.

food can also be viewed as a means of lowering water activity, since water in the form of ice is unavailable to microorganisms.

Although the lowering of water activity is a widely-used method for inhibiting microbial growth, it must never be seen as a means of killing. Although some vegetative cells may die as the water activity is lowered, many survive, and spores are particularly resistant. Furthermore, viable pathogenic bacteria such as salmonellae can often be detected in egg powder or milk powder. Thus, lowering of water activity must be viewed only as a means of inhibiting growth.

4.5   ATMOSPHERIC COMPOSITION

Microorganisms are divided into four groups, depending on their requirements for oxygen:

- aerobes: these organisms cannot grow in the absence of oxygen
- microaerophiles: these organisms cannot grow in the complete absence of oxygen, but they will grow at concentrations much lower than those found normally in the atmosphere
- obligate anaerobes: these organisms cannot grow in the presence of oxygen. Oxygen is toxic, and, in some cases, will kill these organisms
- facultative anaerobes: these organisms can grow in either the presence or absence of oxygen, by adjusting their metabolism appropriately.

Whereas all fungi and green algae are aerobic, different families of bacteria can be found in all of these four groups.

Control of microbial growth on food can often be achieved by manipulation of the gaseous atmosphere surrounding the food. For example, during the canning of food, oxygen is removed from the can, so that if microbial spoilage does occur it will be due to anaerobic rather than aerobic microorganisms.

For non-canned foods which have a high water activity, e.g. meat or cheese, most of the spoilage organisms are aerobic. Hence, if oxygen can be removed from the immediate surroundings of the food, it should be possible to inhibit the growth of the important spoilage organisms, and so achieve a longer shelf-life for the food. This is the basis of **vacuum packaging** and **modified-atmosphere packaging** of food in plastic packages which are impermeable to gases.

The technique of vacuum packaging is used to inhibit bacterial growth and to extend the shelf-life of food. Normally, aerobic organisms will grow on the surface of the food, even at chiller temperatures, and spoilage will occur when the bacterial numbers reach about $10^8/cm^2$. In vacuum packaging, the food is wrapped in a gas-impermeable wrapping and the package is evacuated. Any remaining oxygen will be rapidly depleted by respiration of the food (e.g. in fresh meat) or by respiration of aerobic bacteria. Once the oxygen is exhausted, anaerobic bacteria commence growth. However, growth without oxygen is less efficient, and so growth is slower and the bacterial numbers reach only about $10^7/cm^2$. This means that the spoilage process is delayed considerably. For example, fresh meat stored at a temperature of 0°C has a shelf-life of only 7 days; vacuum packaged meat, however, stored at the same temperature, has a shelf-life of 12–15 weeks.

In modified atmosphere packaging, the food is wrapped in gas-impermeable wrapping, followed by removal of the oxygen by evacuation and replacement with some other gas mixture. For example, gas mixtures containing up to 10% of carbon dioxide, but no oxygen, can be used to inhibit the growth of most spoilage microorganisms. This system can be used with chilled meat products, and also to extend the storage life of perishable bulk packed dried foods such as whole milk powder.

4.6   RADIATION

Visible light has no effect on non-photosynthetic organisms, such as those encountered in the food industry. However, ultra-violet (UV) light and atomic radiations such as x-rays and gamma-rays can kill microorganisms.

4.6.1   Ultraviolet Light (UV)

Ordinary sunlight contains UV light, but most of it is filtered out by ozone before it reaches the ground. Also, it is readily absorbed by glass. It kills bacteria by initiating reactions in the cellular DNA, thus preventing cell reproduction and causing death.

If UV light is directed onto the surface of food, it can kill the bacteria on the surface. However, this is not a practical preservative technique in industry because:

- UV light often cannot penetrate through packaging material
- the intensity of UV light at a given point is inversely proportional to the square of the distance. Hence, the light source must be placed close to the food surface
- UV light can contribute to rancidity.

Although UV light is rarely used for food preservation, it is sometimes used to ensure low bacterial counts in potable water. In this case, the water passes very close to the light source, and it acts as a final treatment method prior to using the water as a food ingredient, e.g. in the brewing industry, the water used for diluting fermenting wort is sometimes treated in this way.

UV lights are sometimes used to sterilise small rooms, as used in microbiology laboratories. They are reasonably effective, but only for small areas. Operators should not be present when the lights are in use because of their danger to eyesight, and their carcinogenic effect on skin.

4.6.2   Atomic Radiations

Atomic radiations such as x-rays and gamma-rays are extremely lethal to microorganisms (and dangerous to personnel!). The technique of **food irradiation** using gamma-rays is now used in many countries for certain foods, but its application is the subject of much controversy. In some countries, e.g. New Zealand, food irradiation remains banned.

The ionising radiation used in food irradiation is of two main types; either it is from large x-ray machines, or it is from a radio-isotope source such as cobalt 60. Irradiation rapidly kills the bacteria on food surfaces, e.g. poultry, and can also be used to sterilise dried, packaged foods. Hence, it can be used to reduce the levels of spoilage and food poisoning organisms, and extend the shelf-life of the product. Although there is no doubt regarding the effectiveness of irradiation as a food preservation technique, its opponents object on the following grounds:

- some bacterial spores are not killed by irradiation. Hence, the killing of harmless competing bacteria could allow the rapid growth of pathogens such as some clostridia
- the technique of irradiation of the final product could lead to deterioration in standards of hygiene and sanitation during processing
- the radiation may alter the chemical structure of the food, and may cause flavour changes and nutrient depletion
- foods that have been spoiled may be irradiated and sold as if they were fresh.

No doubt, this controversy will continue for many years, but use of the technique is likely to increase in the future.

4.7   REMOVAL OF MICROORGANISMS BY FILTRATION

Filtration does not kill or inhibit the growth of bacteria. Rather, it is used to physically remove microorganisms from liquids, e.g. foods, and gases, e.g. air supplies.

4.7.1   From Liquids

To sterilise liquids, heat is usually used. Some liquids, however, are adversely affected by heat, and so some other technique must be applied. **Membrane filtration** is often used to sterilise such thermolabile liquids.

A membrane filter is made of cellulose acetate or nitrate, or materials such as polyvinylidene difluoride. The filter is effective in filtering out microorganisms on the basis of its pore size, e.g. if the pore size is 0.22 μm, then particles larger than this cannot pass through the filter (Figure 4.8). In this way, all microorgan-

Figure 4.8   Diagrammatic representation of a membrane filter. The bacterial cells are too large to pass through the pores of the filter, and so they are removed from the liquid.

isms can be removed from the liquid. The technique is used in the wine industry to sterilise the product prior to bottling.

A disadvantage of the membrane filter is that if the load of particles, including microbial cells, is too high, the filter rapidly blocks and must be replaced. To avoid this, the liquid can be pre-filtered through a diatomaceous earth (Kieselguhr) filter prior to membrane filtration. This will remove most of the particles, leaving only a small load for the membrane filter. While membrane filtration can guarantee sterility, this must never be assumed with diatomaceous earth filtration. The technique is used in the brewing industry, where it removes colloidal protein material and yeast cells from beer. Diatomaceous earth is added to the beer as it passes into the filtration unit. Yeast cells and other particles become adsorbed to the powder, which is retained within the filtration unit by means of a fine mesh. Thus, the beer leaving the unit is relatively free of all particles.

4.7.2   From Gases

Membrane filtration is used to remove microorganisms from gases, in much the same way as it is used for liquids. The technique is widely used to remove particles from the air supply to a food processing area. Air is pre-filtered through, e.g. a fibreglass filter, prior to the membrane filter. This system is known as a high efficiency particulate air (HEPA) filter. In addition to removing microorganisms, the technique removes dust, odour and smoke from the air supply.

Air filters should be inspected regularly to ensure that they are functioning correctly. If they become blocked, they should be changed, and they must be kept dry. Removal of microorganisms and dust from the air entering a food processing department is an integral part of sanitation, in that it removes a potential source of contamination of the product.

Chapter 5

FOOD-BORNE ILLNESSES AND FOOD SPOILAGE

The objective of the food industry is to produce food that is wholesome, fresh and clean. In this chapter we shall discuss the various illnesses that can be transmitted by food, and some of the major microbial causes of food spoilage. Although food-borne illnesses can be caused by microorganisms or by chemicals (e.g. residues of cleaning agents, pesticides), emphasis will be placed on the former since they are much more common in practice. Microbial food-borne illnesses are of three major types:

- **food-borne diseases** — these are caused by microorganisms which never, or only rarely, grow in food. Hence, their control relies on, first, preventing them from gaining access to the food, and, secondly, killing them or removing them from the food. They cannot be controlled by measures which simply inhibit the growth of microorganisms, e.g. refrigeration;

- **infection-type food poisoning** — these are illnesses caused by microorganisms which have contaminated the food, and subsequently have multiplied sufficiently to provide an infective dose. The organisms then reproduce in the gastrointestinal tract and produce the characteristic symptoms. Their control relies on a combination of prevention of contamina-

49

tion, inhibition of growth, and killing them or removing them from the food;

- **intoxication-type food poisoning** — these are illnesses which are caused by microorganisms growing in or on the food, and producing a chemical which is toxic to the consumer. This toxin interacts with the gastrointestinal tract such that the characteristic symptoms are observed. Control relies mainly on prevention of contamination and prevention of microbial growth and toxin production in the food. Killing of the organisms may be of little use if the toxin has already been produced in the food.

5.1   INFECTION-TYPE FOOD POISONING

5.1.1   *Salmonella*

This is a Gram negative facultative anaerobe which is found naturally in the intestines of animals, birds and humans. There are over 2000 different serotypes (strains), all of which are considered to be pathogens, although in any particular country only about 200 serotypes are detected, and perhaps as few as 10 contribute significantly to human disease.

The major source in food is faecal contamination, thus meat, particularly poultry and eggs are widely implicated in outbreaks of salmonella. However, general lack of hygiene and cross-contamination can lead to other foods also being involved.

Salmonella can usually be killed by pasteurisation and by adequate cooking in the household. It can grow at temperatures down to 5°C, and can survive in dried and frozen foods. Thus, it is occasionally found in dried egg and milk products. It is sensitive to low pH values.

There are two principal types of salmonella infections in humans:

- enteric fever (typhoid and paratyphoid), caused by *S. typhi* and *S. paratyphi*. These species are very host-specific, and infect only humans. Hence, a human "carrier" is almost always involved in outbreaks of the disease. In developed countries, where there are reliable sources of drinking water, food is the most common mode of transmission. In developing countries, however, water is the major mode;

- salmonella enteritis (food poisoning), caused by most serotypes. After ingestion, the microorganisms multiply in the intestines, and diarrhoea occurs 8–24 hours later. Other symptoms include nausea, vomiting and abdominal pain, and possibly, chills, headache and muscular weakness. In healthy adults, the symptoms persist for just a few days but in old or sick people, dehydration, collapse and death can occur. Up to 5% of patients become "symptomless carriers", excreting *Salmonella* in their faeces for several months. These carriers represent a potential source of contamination and should not be employed in areas where food or its packaging material is handled.

The infective dose of *Salmonella* varies with the serotype. In some cases, 5 cells per 100 grams of food may be sufficient to initiate an infection, while in other cases, ingestion of 10000 cells is required. The regulatory requirement in processed food is generally for zero in a specified quantity.

5.1.2 *Listeria monocytogenes*

There are several species of *Listeria*, but only *L. monocytogenes* is considered to be pathogenic. It is found widely in nature and in the intestinal tract of animals, birds and humans, and has been isolated from many raw and processed foods. Although it has been known as a pathogen for many years, it is only recently that food has been identified as a vehicle of infection.

The organism is a Gram positive facultative anaerobe which can readily be killed at pasteurisation temperatures. However, its most significant property is its ability to grow at temperatures as low as 3°C, i.e. it is psychrotrophic. For this reason, *L. monocytogenes* appears to be a particular problem in "ready-to-eat" foods, which have been cooked and then stored chilled prior to sale to the consumer. If the heat treatment has been insufficient to kill the organism, multiplication can readily occur during storage.

The infective dose appears to be as low as 100–1000 viable cells, and the incubation period in adults can be up to several weeks. Most healthy adults exhibit little or no symptoms. However, certain people, e.g. pregnant women and immunodeficient people, can exhibit severe flu-like symptoms, and there is a relatively high fatality rate. Pregnant women may have only mild symptoms, but bacteria can infect the foetus causing miscarriage, stillbirth or illness in the newborn baby.

Up to 1% of the human population are carriers of *L. monocytogenes*, and thus are sources of contamination.

5.1.3 Pathogenic *E. coli* and *Shigella*

Escherichia coli and *Shigella* show biochemical differences from each other, but they cannot be distinguished by the technique of DNA hybridization. This means that their DNA compositions are very similar, and so some authorities treat them as one group.

They are Gram negative, facultative anaerobes, some strains of which can grow at 3°C. *Shigella* has long been recognised to be pathogenic. Some strains are very virulent with an infective dose of as low as 100 cells. The major source is the human gastrointestinal tract.

E. coli has been used as an indicator organism in foods for many years, its presence indicating that faecal contamination (from animals or humans) has occurred, and that pathogens such as *Salmonella* might be present. However, it is now recognised to be a pathogen in its own right, and the food industry should recognize this fact. For healthy adults, a very high (10 million cells) infective dose is required, but a much lower dose can cause illness in the young and the sick. The major symptom is diarrhoea, and *E. coli* appears to be the causative agent for "travellers' diarrhoea" as well as for "infant diarrhoea" and "bloody diarrhoea".

Both *E. coli* and *Shigella* are readily destroyed at pasteurisation temperatures, and are sensitive to low pH values.

5.1.4   *Yersinia enterocolitica*

Some strains of this bacterium can cause infections characterised by diarrhoea, fever and abdominal pain. Young children are much more susceptible than adults. The infective dose is at least 10 million cells per gram of food, and the disease is not usually fatal.

The organism is a Gram negative facultative anaerobe which can grow at temperatures as low as 0°C. It is readily killed by pasteurisation, and is sensitive to low pH values.

The major source is the gastrointestinal tract of animals, especially pigs. Hence, the foods of concern include meat, particularly pork, poultry, raw milk and shellfish.

5.1.5   *Vibrio*

V. parahaemolyticus is a Gram negative, facultative anaerobe with a minimum growth temperature of 5°C. Not all strains are virulent and the infective dose is very high. After ingestion, there is an incubation period of 12–24 hours, followed by diarrhoea with severe abdominal pain. The symptoms persist for 3–4 days, but the disease is not usually fatal.

The bacterium is a moderate halophile, so it grows well at relatively high salt concentrations, It is found naturally in warm coastal waters, where it contaminates fish and shellfish. These foods, particularly when eaten raw, are the major cause of the food-borne illness, although there can be cross-contamination to other foods.

V. cholerae is the cause of epidemics of cholera, an explosive diarrhoea that can kill up to 50% of infected individuals, if left untreated. Often, the infection is waterborne, particularly in developing countries, where drinking water is contaminated with faeces. In canned products, the bacterium can enter the can if cooling is performed using untreated water. Food-borne infections are often associated with shellfish or salad vegetables which have been washed in untreated water. Although multiplication of *V. cholerae* in food or water is not necessary for infection to occur, the organism does grow well in a variety of foods, and so is classed as an infection-type food poisoning.

5.1.6 *Clostridium perfringens*

This is a Gram negative anaerobic bacterium which is capable of forming resistant endospores. Although the vegetative cells can be readily destroyed at pasteurisation temperatures, the spores can withstand boiling. The minimum temperature for growth of *C. perfringens* is 15°C.

This type of food poisoning can be quite common, but because of its relatively mild symptoms it currently receives less attention than those poisonings which can be life-threatening. A relatively large number of viable cells must be ingested for symptoms to occur. After 10–12 hours incubation, the cells produce a toxin as they sporulate in the gastrointestinal tract, causing abdominal pain and diarrhoea which last for about 12 hours.

C. perfringens is found widely in nature, in soil and in faeces. Consequently, it is often found on meat. The disease is almost always associated with a meat dish that has received insufficient cooking to destroy the resistant endospores. After cooking, if the dish is cooled and then stored at room temperature for some hours prior to consumption, the spores can germinate and multiply to give large numbers of vegetative cells. Prevention is by storage at refrigeration temperatures. If the food is to be reheated before consumption, it must be done thoroughly enough to destroy all viable bacteria.

5.1.7 *Aeromonas*

The importance of this bacterium is not really clear, but it may be an opportunistic pathogen, causing diarrhoea in young children. It is a Gram negative facultative anaerobe which can grow at temperatures down to 3°C. It is sensitive to pasteurisation and to low pH values.

It is found widely in faeces from animals and humans, and is most associated with such foods as meat, poultry and shellfish.

5.2 INTOXICATION-TYPE FOOD POISONING

5.2.1 Botulism

This is caused by the bacterium *Clostridium botulinum*. Two types are important: Type B is commonly found in soil, and is associated with foods such as meat and vegetables. Type E is of marine origin, and is associated with fish and other marine products. In the USA, home-canned or home-bottled vegetables are the major causes of botulism, while in Japan they are raw and fermented fish products.

C. botulinum is a Gram positive, anaerobic spore-former which can grow at temperatures as low as 4°C. The vegetative cells are killed by pasteurisation, but the spores can survive boiling for up to six hours. The vegetative cells are sensitive to low pH values and to low water activity.

The disease symptoms are caused by the toxin which is produced when vegetative cells are growing on food. The toxin is extremely potent and fatal, although it can be destroyed by heating at 85°C for 15 minutes. Although botulism is relatively uncommon, it is so serious that for many years, destruction of *C. botulinum* has been a major theme in food processing, particularly with canned products. Since the bacterium can grow in chilled foods, control often depends on other environmental parameters such as pH and water activity. However, with the increasing demand from consumers for a reduction in the salt content of foods, and with restrictions in the use of chemical food preservatives, there is a danger of losing inhibitory control of this lethal microorganism.

The incubation time for botulism is 12–36 hours after ingestion of the toxin. The first symptom is acute digestive disturbance followed by nausea and vomiting, and possibly diarrhoea, together with fatigue, dizziness and headache. Involuntary muscles become paralysed, spreading to the respiratory system and heart. Death results within 3–6 days, usually from respiratory failure.

5.2.2   Staphylococcal Food Poisoning

Strains of staphylococci are designated as either coagulase positive or coagulase negative, depending on their ability to coagulate blood serum. Coagulase positive strains (mainly *S. aureus*) are often pathogenic, and are associated with septic wounds and boils. Some of these pathogenic strains also produce a toxin which, when ingested, can produce the characteristic symptoms of staphylococcal food poisoning. The organism is a Gram positive facultative anaerobe, whose minimum growth temperature is 7°C, although the lowest temperature for toxin production is 10°C. It is killed by pasteurisation but can grow at levels of water activity as low as 0.84. Of particular importance is the fact that the toxin can remain potent even after boiling for 10 minutes.

Staphylococcal food poisoning is one of the major causes of foodborne illness, although it is only rarely fatal. As many as 10^6 cells per gram of food may be necessary for sufficient toxin to be produced to cause illness. The incubation time may be as short as one hour, followed by nausea, vomiting, retching and diarrhoea. Headache and chills may occur, but the duration rarely exceeds two days.

The principal source of *S. aureus* is the food handler. Up to 50% of healthy people are "carriers", usually in the nose, from where it can be transferred to the hands and, consequently, to food. Hence, the presence of *S. aureus* on food is often a reflection of the personal hygiene of the food handler. The other main source is milk from an animal suffering from staphylococcal mastitis. Cheese made from such milk may contain sufficient toxin to produce the disease symptoms.

5.2.3 *Bacillus cereus*

Although several species of *Bacillus* such as *B. subtilis* and *B. licheniformis*, have been implicated in outbreaks of food poisoning, *B. cereus* is, by far, the most well-known. It is a Gram positive, facultatively anaerobic, spore former, with a minimum growth temperature of 5°C. Pasteurisation kills the vegetative cells, but not the spores.

The major sources of the organism are grains (particularly rice), flour, starch, spices and animal products contaminated with soil. Up to 100 000 cells per gram of food may be required to produce sufficient toxin to cause illness. Two types of toxin are produced, which can cause two different types of illness:

- an emetic syndrome, with nausea and vomiting 1–5 hours after consumption, usually involving a rice dish
- a diarrhoeal syndrome, with an incubation period of 8–16 hours. This usually involves meat dishes where large amounts of spices are added late in cooking. The spores survive, and germinate on cooling.

5.2.4 Mycotoxins

Many fungi produce a range of chemical compounds, some of which may be toxic to humans and to other animals. The best known examples are the aflatoxins, which are believed to be carcinogenic, but there are many others which can cause a variety of symptoms. The fungi may produce the toxins during growth on crops in the field, or during storage of the food product. The main problem appears to be with stored grains, resulting not only in a health hazard, but also in economic loss. The solution to the problem is in improved storage procedures, paying particular attention to factors such as temperature and moisture content of the grain, so that fungal growth is minimised.

5.2.5   Algal Toxins

In recent years, there has been increasing incidence of food-borne illness which is caused by toxins produced by algae. These outbreaks are almost always associated with "algal blooms" or "red tides", when climatic conditions and sea temperatures, coupled with good nutrient availability, result in enormous growth of certain algae. These algae form a source of food for shellfish, which are filter-feeders and so concentrate the algae within themselves. The algae contain toxins which, if the shell-fish are consumed by humans, can cause a variety of symptoms such as vomiting, but which are rarely fatal. Prevention of this type of illness will rely on non-consumption of shellfish which have been harvested from areas known to contain high levels of algae.

5.3   FOOD-BORNE DISEASES

5.3.1   *Campylobacter*

This is becoming increasingly recognised as one of the major causes of infectious diarrhoea in humans, at least in developed countries. The symptoms include profuse watery diarrhoea, or bloody diarrhoea, associated with severe abdominal pain. The symptoms can last for several weeks, and be very debilitating, but the disease is rarely fatal.

The bacterium is a Gram negative microaerophile which does not grow at temperatures below 30°C, and which is easily killed by dehydration and pasteurisation temperatures. However, it will survive at temperatures down to 4°C. Several species may be involved but the major one is *C. jejuni*. The major source is the gut of poultry, cattle, pigs, sheep and a variety of wild animals and birds. Infections are believed to arise from meat infected with gut contents at the time of slaughter. If the meat surface is allowed to dry, as with red meats, the numbers of Campylobacter decrease markedly. However, surface-drying does not occur with chickens. The infective dose is believed to be less than one thousand cells.

5.3.2   Viruses

Many viruses can cause gastrointestinal diseases. Transmission may occur by an aerosol or by direct contact with an infected person, but can also be by food-handlers and contaminated water. Filter-feeding shellfish, harvested from water contaminated with human sewage, can be a major source of infection.

5.3.3   Protozoa

Several protozoa, such as *Giardia* and *Entamoeba*, can cause gastrointestinal illness. They are often transmitted by water, but food can also be responsible. The major source is the gut of animals. Some species have resistant spores, or cysts, which can survive chlorination and remain viable in water for long periods of time.

5.4   FOOD SPOILAGE

Spoilage of food is considered to be decay or decomposition of an undesirable nature. The food is so changed that, **for aesthetic reasons** (e.g. odour, appearance) it is undesirable to eat.
   Spoilage can be due to one or more of the following:

- growth and activity of microorganisms
- insects
- action of the food's own enzymes
- chemical reactions in the food
- physical changes in the food, such as those caused by freezing or drying.

Spoilage by microorganisms is by far the most common, and will be discussed in most detail here.

Different types of microorganisms cause different types of spoilage. The type of spoilage that occurs in any situation depends on:

- the types and numbers of organisms present initially on the food
- the **intrinsic factors** of the food, i.e. the nutrients present, the pH, and the water activity
- the **extrinsic factors** of the food, i.e. the environmental factors such as temperature and atmospheric composition.

Most raw foods have a variety of microorganisms on them. The particular organisms that will grow will depend on the intrinsic and extrinsic factors of the food. Usually, a single type of organism will dominate, and cause a characteristic spoilage.

There are two main sources of contamination of food: natural contamination and contamination during processing and/or storage.

5.4.1   Natural Sources of Contamination

Animals and plants have a typical population of microorganisms on their surfaces. In addition animals have an intestinal population, but the inner tissues of both plants and animals contain relatively few organisms.

Plants and animals are often contaminated from soil, sewage, water and air. Soil contains a very wide variety of organisms, and surface washing of foods to remove these is an integral part of many food processing procedures. Soil is often contaminated with faeces so that the hides and feet of meat animals are often badly contaminated from this source.

Untreated sewage can cause heavy contamination of plants if it is used for irrigation purposes. Natural waters contaminated with sewage can be a source of contamination for fish and for animals which drink the water. Shellfish are particularly prone to contamination since they are filter-feeders, and thereby concentrate organisms within themselves.

5.4.2 Contamination During Processing

Contamination sources during processing include:

- ingredients (including intestinal contents of meat animals)
- insects, rodents
- people
- air, dust
- dirty equipment and surroundings
- water.

Control of contamination from these sources is discussed elsewhere in this book.

5.4.3 The Spoilage Process

For any particular food containing a population of different microorganisms, the intrinsic and extrinsic factors will determine which particular one will grow. These factors interact strongly with each other so that any small change in any one of them can markedly change the type of spoilage that occurs. For this reason, it is difficult to generalise about the spoilage process. Nevertheless, the following very simplified points can be made.

5.4.3.1 Water activity

Yeasts and moulds usually grow at lower values than do bacteria. Hence, foods such as honey and bread are spoiled by the former, while moist foods such as meat are spoiled by the latter. Exceptions include salt-tolerant and halophilic bacteria which often grow on salted meat and fish.

5.4.3.2 pH

Yeasts and moulds grow at lower pH values than do bacteria. Thus, acidic foods such as fruit juices are spoiled by the former rather than the latter.

5.4.3.3 Temperature

Below –2°C, any spoilage will be due to psychrophilic organisms, usually fungi. Above this temperature and up to 15°C, psychrotrophic bacteria, e.g. Pseudomonads, generally predominate, while at higher temperatures mesophiles such as coliforms or lactic acid bacteria may become the dominant organisms.

5.4.3.4 Atmospheric composition

At normal oxygen concentrations (i.e. 21% of air), aerobic bacteria such as Pseudomonads or *Bacillus* often predominate. Under anaerobic conditions, however, *Clostridium* and *Lactobacillus* are usually the organisms responsible for spoilage.

5.4.3.5 Nutrient composition of the food

This is obviously important in that to be able to grow, microorganisms must be able to utilise the nutrients present in the food. In general terms, simple sugars and amino acids can be used by a wide range of organisms. However, as the nutrients become more complex, e.g. proteins and polysaccharides, increasingly fewer organisms can use them. Pseudomonads, *Bacillus* and *Clostridium* can utilise a very wide range of nutrients, while coliforms, *Lactobacillus* and *Streptococcus* have a much more limited range, usually being restricted to sugars and amino acids.

Chapter 6

CLEANING AND
SANITISING CHEMICALS

Cleaning, or the removal of unwanted material (dirt, "soil") from equipment and surroundings, is one of the most common operations in the food industry, and one of the most important. Unless equipment and surroundings are properly clean they will offer an ideal environment for microbial growth, which may be transferred to the product, and may attract rodents and insects.

The **objectives** of a cleaning programme are:

- removal of all soiling matter (dirt)
- removal of all pathogens
- removal of all spoilage organisms
- achievement of a low bacterial count on all equipment and surfaces, i.e. less than $10^2/cm^2$.

It is unlikely that equipment surfaces will have a zero bacterial count after cleaning, but provided all dirt (nutrients) has been removed, counts are unlikely to show a subsequent increase. If cleaning, or removal of soiling matter, were not performed bacterial growth would occur and provide a major source of contamination for the food product during the next day's operations.

For a cleaning programme to be successful, the following must be available:

- an adequate supply of good quality (potable) water
- the correct detergent for the job
- the correct sanitiser (disinfectant) for the job
- an appropriate cleaning procedure.

The purpose of the detergent is to remove dirt.

The purpose of the sanitiser (disinfectant) is to kill microorganisms.

The use of a detergent to remove dirt often removes many microorganisms purely mechanically, but the additional use of a sanitiser, if used correctly, will guarantee killing of microorganisms.

6.1   WATER

Water for use as a food ingredient, or for cleaning purposes, must be of a **potable** standard, i.e. it must meet the appropriate standards for chemical and microbiological quality. Methods of water treatment and examples of standards are covered in Chapter 8, and will not be discussed further here.

6.2   DETERGENTS

Detergents assist in the removal of dirt. Water alone is not very effective in removing dirt unless a lot of work is done. This is because of its high surface tension which prevents intimate contact with dirt and equipment surfaces. When detergents are dissolved in water, they reduce the amount of work to be done, and so they assist the cleaning process. They do this by:

- lowering the surface tension, thus thoroughly wetting the surface (Figure 6.1). Chemicals which possess this property are termed **surface active agents** or **surfactants**;

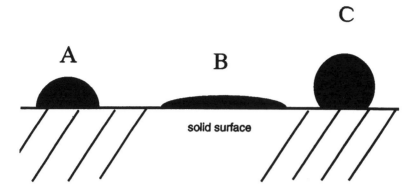

Figure 6.1 Surface tension of liquids. Liquid A is water, B is a surfactant solution, and C is mercury.

Water has a moderately high surface tension and forms large droplets on a solid surface due to the forces within the liquid. When dissolved in water, surfactants lower the surface tension, causing the droplet to collapse and make a far greater contact with the solid surface. In contrast, mercury has a very high surface tension, and forms large globules which can easily roll across a surface, but the area of contact with the surface is very low.

- penetrating and softening the dirt, making it easier to remove by hoses or scrubbing;
- suspending and dissolving solids so that they may be more easily carried away by the water.

In most cases, it is unwise to rely on the detergent solution alone to remove dirt. There is still a need to provide some mechanical energy, such as use of a scrubbing brush or high pressure hose, but the amount required will be much less than if water alone was used.

A good detergent should have the following properties:

- quickly and completely soluble in water
- ability to lower the surface tension of water, and so thoroughly "wet" a surface

- ability to dissolve and/or suspend ("soften") dirt, including the ability to emulsify fats
- ability to cope with hard water, i.e. prevent deposition of insoluble calcium and magnesium salts ("sequestration" properties)
- ability to remove scale and rust
- non-corrosive to equipment or structural surfaces, and safe for personnel to use
- good rinsing properties, to prevent the redeposition of suspended dirt
- odourless, so that no lingering odours can taint the food product.

Unfortunately, there is no single chemical compound that meets all these specifications. Hence, most of the commercial detergents that are available are **mixtures (formulations)** of two or more chemicals.

Soap (sodium/potassium stearate or palmitate) is no longer widely used in the food industry because of its lack of effectiveness, particularly under acid conditions or when hard water is used. Its main property is to lower the surface tension of water, i.e. it acts as a surfactant. However, in some situations, it is still the best cleaning agent, e.g. liquid soap for handwashing.

6.2.1   Ingredients of Commercial Detergents

Commercial detergents are composed of two or more ingredients, each of which contributes a particular property. These ingredients are described below.

6.2.1.1 Alkalis

These have dissolving and emulsifying properties on food solids, particularly proteins and fats.

Sodium hydroxide (caustic soda) is the cheapest and strongest alkali. It is a powerful detergent and is used to dissolve protein, and to convert oils and fats to soaps. However, it has no buffering action, is very corrosive to many surfaces, and precipitates calcium and magnesium salts (mineral hardness) from water. Because of its corrosive action it is not recommended for cleaning some metal equipment, and it can be hazardous to personnel using it.

Sodium metasilicate is an effective cleaner for many purposes, and possesses anti-corrosive properties. It should not be used at water temperatures greater than 62°C, or else the soil-detergent mixture may precipitate. It is an excellent emulsifying and suspending agent, and has good rinsing and wetting properties. Although less effective as a cleaner than sodium hydroxide, sodium metasilicate is less corrosive and safer to personnel.

Sodium carbonate is less effective than either sodium hydroxide or sodium metasilicate, and is rarely used in modern detergent formulations.

Detergents based on alkalis, so-called **alkaline detergents**, are widely used in the food industry. An important factor concerning them is their active alkalinity; this gives an indication of usefulness in cleaning, and also of corrosion dangers. The higher the active alkalinity, the greater the effectiveness and corrosivity.

6.2.1.2 Phosphates (polyphosphates; sequestering agents)

These prevent the formation of insoluble salts, formed by interaction of hard water with the detergent. Hence, they prevent the formation of "scale" which can cause equipment to become dull in appearance, and they improve the ease with which the detergent-soil suspension can be rinsed off metal surfaces.

Manufacturers can use several different phosphates in commercial detergents. **Tetrasodium pyrophosphate** is the most widely used, but it is less effective than others. Although it is stable in hot alkaline solution, it is rather slow to dissolve, and detergents containing this compound should be dissolved in hot water. **Sodium tripolyphosphate** and **tetraphosphate** are both superior to pyrophosphate in terms of sequestering power on calcium hardness, and both are readily soluble in warm water. **Sodium hexametaphosphate** is the most effective agent for prevention of scale deposition, but it is unstable in hot alkaline solutions.

In recent years, some opposition has developed to the use of phosphates in commercial detergents. This is due to excessive amounts of phosphates reaching waterways, particularly lakes, where they contribute to rapid growth of algae (algal blooms) which subsequently die and decay, creating bad odours and depleting the waterway of oxygen. This process is known as **eutrophication**. However, the major source of phosphate in waterways is often agricultural fertilizer rather than commercial detergents. Possible substitutes for phosphate in detergents include sodium citrate and sodium gluconate, but neither of these is as cost-effective as phosphate.

6.2.1.3 Synthetic surfactants

These were developed during the 1950s as synthetic substitutes for soap. One of their main properties is the ability to lower the surface tension of water, and thus thoroughly "wet" a surface and penetrate the dirt. This, in turn, allows the other ingredients in the detergent formulation to perform their role more effectively. Several types are available.

- **Anionic surfactants** are the most widely used in detergent formulations. They are termed anionic because, when dissolved in water, they give negatively charged ions. This is similar to soap (e.g. sodium stearate) although the anionic group is different. An example is sodium lauryl sulfate, $C_{12}H_{25}OSO_3^-Na^+$. (Sodium stearate is $C_{17}H_{35}COO^-Na^+$.)

These surfactants are effective in cleaning partly because of their emulsifying properties on grease and fats. The long carbon chain (hydrophobic group) associates with the fat while the anionic (hydrophilic) group associates with water. This breaks up the fat into minute droplets and assists its removal (Figure 6.2).

- **Non-ionic surfactants** are particularly used when foam generation is required. Sometimes they are used in combination with anionic surfactants but sometimes they are used

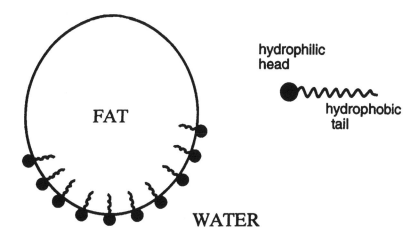

Figure 6.2 Fat globule surrounded by surfactant molecules. The long hydrophobic tail of the surfactant molecule associates with the fat, while the hydrophilic head associates with water.

alone, though in combination with other detergent ingredients. When dissolved in water, they do not form ions, but they act in a similar manner to the anionic type in that they possess a long hydrophobic carbon chain and a hydrophilic group. An example is an alkyl polyoxyethylene alcohol, $C_{14}H_{29} - O - (C_2H_4O)_n - C_2H_4OH$.

The term "biodegradable" is often used in connection with commercial detergents, and usually refers to the surfactant which is present. When first manufactured during the 1950s the hydrophobic carbon chains of these molecules were usually branched rather than linear. Because of this, they were degraded only slowly in waste treatment plants and waterways, and often caused unsightly foam. Nowadays the manufacturing process produces linear carbon chains which are degraded much more readily, and most available surfactants are biodegradable.

- **Cationic surfactants** are poor surfactants, and are not used for this purpose. They are, however, excellent sanitisers, and are discussed in section 6.3.

6.2.1.4 Acids

These are used to remove mineral "scale" (formed from the deposition of calcium salts), rust and other alkali-insoluble material. Acids commonly used include nitric acid and phosphoric acid, both of which can be highly corrosive. Milder products are sulfamic acid, citric acid, sodium bisulfate and a mixture of peracetic acid and hydrogen peroxide. The stronger the acid, the more effective it is in removing dirt, but it is also more corrosive to equipment and dangerous to personnel.

6.2.1.5 Proteolytic enzymes

These are usually produced from microbial sources, e.g. the bacterium *Bacillus subtilis*. Their important attribute is the ability to hydrolyse protein to amino acids, and so they are used to remove protein residues. Their optimum pH for activity is pH 8–9, while their temperature optimum is normally 45–50°C.

6.2.1.6 Abrasives

These are solid materials, such as pumice or volcanic ash, which may be added to commercial detergents as **scourers**, so that the product is a powder or paste. Combined with scrubbing, these powders are very effective in removing stubborn dirt from surfaces, but they should be used with care to avoid scratching surfaces or leaving behind powder material which may subsequently contaminate the food product.

In commercial detergents, two or more of the above ingredients are mixed to give a formulation suitable for use in the food industry.

6.2.2   Commercial Detergents

Commercial detergents may be divided into the following groups.

6.2.2.1 Alkaline detergents

These generally contain an **alkali**, a **surfactant** and a **phosphate**. There are two types depending on strength:

- **"medium" alkaline detergent**, where the alkali used is generally sodium metasilicate. A typical formulation would be:

sodium metasilicate	80 g
sodium tripolyphosphate	16 g
anionic surfactant	4 g

It would be dissolved in water at about 100 g per 10 litres of water. A typical pH in solution would be pH 10–10.5, with an active alkalinity of < 40%.

This type of detergent is good for removing protein and fat, and is widely used for general cleaning in food factories. It is non-corrosive provided it is used according to the manufacturer's instructions. Its use at too high a concentration could lead to corrosion problems:

- **"heavy duty" alkaline detergent** has a composition similar to the "medium" version, but the alkali used is generally sodium hydroxide. A typical pH in solution would be pH 12–13, with an active alkalinity of > 40%.

This type of detergent is corrosive and dangerous to personnel and should be used with care. It is particularly useful for removing stubborn grease or dirt, and is often used on a weekly basis for this purpose. It should always be used according to the manufacturer's instructions.

6.2.2.2 Acid detergents

These contain an **acid** in combination with a **surfactant**. A typical pH in solution is pH 3. Because of their potential to cause corrosion, a proprietary corrosion inhibitor is often present in the formulation.

Acid detergents are used when mineral deposits, or rust, accumulate on metal surfaces, or when an alkaline detergent has been used for some considerable period of time leaving behind alkali-insoluble deposits. In some circumstances it is necessary to use both an alkaline detergent and an acid detergent in the cleaning procedure; in this case the former is generally used first, followed by thorough rinsing with water prior to application of the acid detergent.

When using acid detergents, because of their corrosion properties it is important to adhere to the manufacturer's instructions. Sulfamic acid and sodium bisulfate are the safest acids to use, but phosphoric acid and nitric acid are better "scale" removers. Peracetic acid, in combination with hydrogen peroxide, is excellent.

6.2.2.3 Neutral detergents

These contain anionic and/or non-ionic **surfactants** at a pH close to 7. They are mainly for dishwashing and household use, and have little application in food processing areas.

6.2.2.4 Proteolytic enzyme formulations

These contain a **proteolytic enzyme**, a **surfactant** (often of the non-ionic type to allow generation of foam) and a small amount of **alkali** to pH 8–9. They are used at a temperature of 45–50°C, and are excellent for protein removal.

Proteolytic enzyme formulations and alkaline detergents are both effective at removing proteins. The enzymes have the advantage of performing optimally at lower temperatures than the alkaline detergents, thus giving a cost/energy saving, but they are less effective in fat removal. Hence, the choice of which to use in a given situation often depends on the amount of fat that is present.

6.2.2.5 Detergent sanitisers

Some detergent formulations are available which, in addition to the usual ingredients, also contain a sanitiser (disinfectant) compound. Such formulations possess both detergent and sanitiser properties. Some manufacturers claim that they provide extra safety from the bacteriological viewpoint, and that they permit cleaning and sanitising in one operation, thus saving time and labour. This may be true for areas which are only lightly soiled, but in situations of heavy soil, or in particularly sensitive process areas, these formulations should be treated as detergents, and a separate sanitiser step should be used (refer Chapter 7).

Three major types of detergent-sanitisers are available:

- alkaline detergent + hypochlorite
- alkaline detergent (**not** containing an anionic surfactant) + quaternary ammonium compound
- acid detergent + iodophor.

In the case of products containing hypochlorite, this chemical also assists in the removal of protein residues.

6.2.3   Choice of Detergent

The type of detergent to use in any situation is determined by:

- the type of dirt to be removed, e.g. an alkaline detergent is used for alkali-soluble dirt (Table 6.1)
- the type of construction materials present, i.e. it is important not to corrode equipment (Table 6.2)
- the type of cleaning technique used, e.g. if much manual labour is used, it is important not to use excessively acid or alkaline detergents.

Within each type of detergent it is often necessary to select which brand to use from the hundreds on the market. Some laboratory tests for preliminary investigations are available (see Appendix 1). However, it must be remembered that only under actual working conditions can a detergent be truly evaluated. It should also be borne in mind that different companies within the same industry, and even different departments within the same factory, may use quite different brands of detergents depending on the particular circumstances. A brand which is claimed by the manufacturers to be satisfactory under all conditions may, in fact,

Table 6.1 Choice of detergent.

Dirt type	Detergent
Fats, grease	Alkaline detergents
Protein (including blood)	Alkaline detergents Enzyme detergents
Mineral deposits, calcium scale, milkstone, rust	Acid detergents

Sugars and carbohydrates dissolve readily in water, so the choice of detergent is governed by the other types of dirt present.

Table 6.2 Cleaning implications of some construction materials.

Material	Consideration
Stainless steel	All alkalis safe Use acids and chlorine with care
Mild steel	All alkalis safe Acids are corrosive
Aluminium	Use only mild alkalis
Nickel alloys	All alkalis safe Use acids with care
Concrete	Alkalis satisfactory Do not use acids
Glass	No restrictions
Fibreglass	Use only mild alkalis or mild acids
Ceramics	No restrictions
Plastics	No restrictions, but take care at high temperatures
Rubber	No restrictions, but take care at high temperatures
Paints	Use only mild alkalis or acids, but take care at high temperatures

give only a mediocre performance. Any detergent selected for use must be approved by the appropriate regulatory authority.

The **cost** of a detergent should always be considered, but it is not necessarily a true guide to efficiency or economy. Labour costs can account for up to 70% of the total cleaning cost. Occasionally, an expensive detergent can reduce the labour cost sufficiently to ensure that the expensive brand is actually the most economical.

6.3 SANITISERS

The purpose of sanitisers is to kill microorganisms. They are chemical compounds which can be sprayed onto equipment and surroundings. After performing their function they are rinsed away, and the surfaces, while not completely free of bacteria, should be "commercially sterile" ($< 10^2$ bacteria per cm^2 of surface).

The way in which chemicals kill bacteria can be very complicated, but in most cases sanitisers act by reacting with and destroying proteins and enzymes within the cell. Before considering the use of sanitisers, some definitions should be noted:

- **disinfection** is the act of destroying microorganisms;
- a **disinfectant** is a chemical agent that kills growing, vegetative cells, but not necessarily spores;
- an **antiseptic** is similar to a disinfectant. The major difference between the two is one of use. Disinfectants are generally used on inanimate objects, whereas antiseptics are used on living tissue. The same chemical can, in fact, be used for both purposes;
- a **sanitiser** is a chemical agent that reduces the microbial population to safe levels as judged by public health requirements. Usually it kills 99.9% of growing microorganisms, but not necessarily spores. The process of disinfection would produce sanitization. However, in the strict sense, to disinfect

is not necessarily to sanitise since sanitation implies a condition of cleanliness which disinfection does not;

- for all **practical purposes**, however, a sanitiser is the same as a disinfectant.

6.3.1 Factors Affecting Killing by Sanitisers

6.3.1.1 The type of sanitiser

Some types kill a wide range of microorganisms, and are known as "broad spectrum", while others kill only a narrow range ("narrow spectrum").

6.3.1.2 The concentration of the sanitiser

In general, the more concentrated the sanitiser, the more rapid is its action. At low concentrations, a sanitiser may be only bacteriostatic, while at higher concentrations it is bactericidal. At extremely low concentrations, some sanitisers may actually be used as a nutrient for microbial growth! With many sanitisers, the relationship between concentration and effectiveness is not linear. Rather, it is exponential, e.g. doubling the concentration does not merely double the effectiveness, but increases it by ten times. However, a further increase in concentration may give no further improvement in effectiveness (Figure 6.3).

Hence, there is an optimum concentration of sanitiser, below which the effectiveness is reduced, and above which there is no further improvement. In fact, depending on the type of sanitiser, use at too high a concentration can cause corrosion problems. Normally, sanitisers should be used at the manufacturer's recommended concentration.

6.3.1.3 The contact time

No sanitiser acts instantaneously. Sufficient contact time must be allowed for the sanitiser to penetrate into the microbial cell and react with its target protein. Different sanitisers require different times, but in all cases the time is affected by:

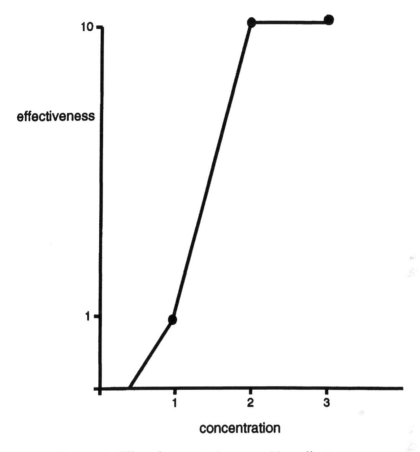

Figure 6.3 Effect of concentration on sanitiser effectiveness.

- **pH**. Extremes of pH (i.e. strong acid or strong alkali) reduce the time required, and thus increase the sanitiser's effectiveness;
- **temperature**. In general, an increase in temperature reduces the time required, although there are exceptions to this. (Ideally, sanitisers are used at room temperature);
- the **concentration** of the sanitiser, as described above.

Most suppliers of sanitisers give a recommended contact time. It is important not to shorten this time, or else the sanitiser will not be as effective as it should be. Conversely, with some sanitisers, too long a contact time may lead to corrosion problems.

6.3.1.4 The concentration of the microorganisms present

In theory, under uniform conditions of temperature, pH, and sanitiser concentration, all cells in a population should be killed at the same time (Figure 6.4a). In practice, however, this rarely happens. Instead, some cells are killed early on, while others survive until much later. This is because the individual cells in the population differ in their susceptibility to the sanitiser. Hence, a graph of the number of survivors plotted against time would be a curve (Figure 6.4b). If a graph is drawn of the logarithm of the number of surviving cells against time, a straight line is obtained (Figure 6.4c). This demonstrates that the death rate is analogous to a first-order reaction, i.e. the number of cells dying is proportional to the number of cells present.

To illustrate this effect, consider the following example (Figure 6.4c):

Under constant conditions, a sanitiser kills 90% of the cells present in 5 minutes.

If :   100 000 (10^5) cells or   100 (10^2) cells are
                                           present initially,

After 5 minutes, the number of cells alive is:
        10 000 (10^4) cells   10 (10^1) cells

After 10 minutes:
        1000 (10^3) cells   1 (10^0) cells

After 15 minutes:
        100 (10^2) cells   0.1 (10^{-1}) cells

Hence, it is clear that the more cells that are present initially, the longer it will take to kill them all. This has major implications

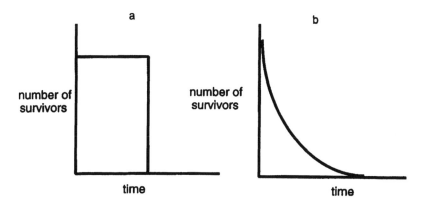

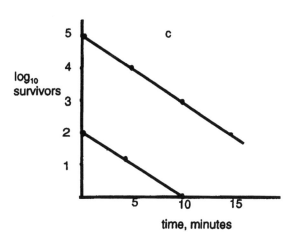

Figure 6.4 (a) Theoretical plot of number of survivors against time during use of a sanitiser. (b) The plot of number of survivors against time that is normally obtained in practice during use of a sanitiser. (c) Plot of logarithm of number of survivors against time, showing that the size of the initial population affects the time taken to kill that population.

for the use of a sanitiser in that for particularly contaminated areas, sufficient contact time must be allowed to kill the large number of microorganisms that are present.

6.3.1.5 The presence of dirt

Most sanitisers act by reacting with cell constituents, particularly protein. However, they are quite general in their affinity for protein, whether part of a living cell or not. Hence the presence of protein or fat can inactivate sanitisers, and prevent them from reaching the target bacteria.

From this, we learn that thorough cleaning must always precede the use of a sanitiser.

6.3.1.6 The surface tension of the sanitiser solution

If the sanitiser is also a surfactant (i.e. it lowers the surface tension of water), its sanitising properties will be enhanced because

- the increased "wetness" gives a better contact with the surface to be sanitised
- surfactants become concentrated on bacterial cell walls, thus achieving a more immediate contact.

Some sanitiser manufacturers mix a surfactant with a sanitiser to achieve the property of increased "wetness".

6.3.2   The "Ideal" Sanitiser

The ideal sanitiser for use in an area where food is processed should have the following properties:

- broad spectrum activity, and organisms should not develop resistance
- readily soluble in water, as all sanitisers are applied in the aqueous phase
- chemically stable on storage, so that sanitising properties are not diminished over a period of several months

- non-toxic to mammals, so that sanitisers can be stored and applied in safety
- resistant to inactivation by organic matter (dirt)
- non-corrosive to construction materials
- surfactant activity
- odourless, to minimise the possibility of product becoming tainted
- available in large quantities at reasonable cost.

Unfortunately, there is no single sanitiser which possesses all of these properties; they all have their particular advantages and disadvantages.

6.3.3   Types of Sanitiser

6.3.3.1 Chlorine compounds

Considered purely from the germicidal viewpoint, halogens (chlorine, bromine, iodine) are the most effective of all sanitisers. They are extremely powerful against virtually all vegetative cells.

Chlorine is a gas, and is widely used in the purification of water supplies. Because it is a gas, however, and highly dangerous to personnel, it is not suitable for use as a sanitiser in the food industry. Hence, what we use in this situation is a "chlorine-containing compound" (chlorophor) which, when dissolved in water, releases chlorine to act on the microorganisms. A disadvantage of these compounds is that they are inherently unstable, but they permit the safe and convenient use of chlorine.

- **Hypochlorites** are examples of such compounds. Calcium hypochlorite ($Ca(OCl)_2$, a powder) and sodium hypochlorite (NaOCl, a liquid) are used both industrially and domestically. Under appropriate conditions they break down and release chlorine, which stays in solution and forms **hypochlorous acid** (HClO) which is the effective germicide:

$$Cl_2 + H_2O \rightleftharpoons HClO + HCl$$

The strength of hypochlorite solutions is expressed as their "available chlorine content". This is a measure of the amount of chlorine which can be released from the hypochlorite (or any other chlorophor). The value can be determined in the laboratory by the starch-iodine titration method, i.e. the hypochlorite is reacted with potassium iodide in acid solution, and the iodine released is estimated by titration with thiosulphate using a starch indicator, the amount of iodine released being proportional to the available chlorine content.

Commercial solutions of hypochlorites for use in the food industry usually have an available chlorine content of around 100 000 ppm. For use, they are diluted to 200–500 ppm. Caustic soda (sodium hydroxide, NaOH) is generally also present in commercial hypochlorites, so that the solution pH is alkaline. Because of their potential corrosive nature, the "contact time" of chlorine-based sanitisers should be carefully noted. This is usually between 2 and 10 minutes.

Although the available chlorine content is a measure of the strength of a hypochlorite solution, this is not always a true criterion of sanitiser effectiveness, since much depends on the **rate of release** of the available chlorine. The rate of release is increased by:

- decreased pH. At acid pH values the rate of chlorine release is so fast that the solutions are potentially dangerous to personnel and corrosive to equipment. It is for this reason that hypochlorites are used under alkaline conditions;

- increased temperature. Hypochlorites are used at room temperature, so that the rate of chlorine release can more easily be controlled;

- increased concentration. Strong solutions are less stable than weaker ones.

Hypochlorites are unstable compounds. To minimise their instability during storage, they should be stored in non-metallic containers, in a cool place, in the dark.

As sanitisers, hypochlorites have the disadvantages of being corrosive and of producing a characteristic odour. They are also potentially dangerous to personnel using them, particularly in confined spaces. In addition, they are readily inactivated by dirt. However, when used correctly, i.e. according to the manufacturer's instructions, they are excellent sanitisers.

- **Inorganic chloramines** are sometimes used in detergent-sanitisers. However, they have a slower rate of release of available chlorine than do the hypochlorites, and so they are less effective as sanitisers. Nevertheless, they do have a place in the treatment of water supplies. Members of this group of chemicals are nitrogen trichloride (NCl_3), dichloramine ($NHCl_2$) and monochloramine (NH_2Cl).

- **Organic chloramines** also have a role in detergent-sanitisers, but their rate of release of available chlorine is even lower than that of the inorganic chloramines.

6.3.3.2 Iodine and iodophors

Iodine is the most effective sanitiser of all the halogens. In addition to killing virtually all types of bacteria, it is effective against some spores. Unfortunately, for use in the food industry, iodine suffers from being only slightly soluble in water, it stains equipment a brown colour, it has a strong odour, and it is an irritant to the eyes and nose. For these reasons, it has no use as a sanitiser.

Iodophors are sanitisers in which iodine is loosely combined with a non-ionic surfactant (section 6.2.1) which acts as a solubilizer and carrier of the iodine. As sanitisers, they have all the good properties of iodine, but none of the bad ones. Thus, they permit the safe application of a very powerful chemical.

The strength of iodophor solutions is expressed as available iodine content, and typical concentrations for use are 10–25 ppm. Commercial solutions usually contain some phosphoric acid, so the solution is pH 3–4. This makes them particularly suitable for use in the dairy industry, where the dirt is often acid-soluble.

The advantages of iodophors include:

- they are powerful sanitisers;
- they are odourless, non-irritant, and show low toxicity to mammals;
- they have detergent properties, and so are less affected by dirt than are some other sanitisers;
- they stain dirt, but not equipment. Hence, it is easy to see whether equipment has been properly cleaned or not;
- they lose their brown colour when inactivated by dirt. Hence, it is easy to see if inactivation occurs;
- they are stable on storage.

However, there are also some disadvantages:

- if used negligently, they may be corrosive, cause staining and produce odour.

6.3.3.3 Quaternary ammonium compounds (QACs, Quats)

These are cationic surfactants which are, in fact, poor surfactants but effective sanitisers. They are organically-substituted ammonium compounds, whose main attributes are safety and non-corrosive properties.

$$\begin{bmatrix} R^1 & R^2 \\ & N & \\ R^3 & R^4 \end{bmatrix} X^-$$

X is a halide or similar radical, while the R constituents each represent an alkyl, aralkyl or heterocyclic radical. This basic structure allows for an unlimited variety of compounds, but probably less than a dozen are of any use as sanitisers. Quaternary ammonium compounds are useful sanitisers because:

- they have good bactericidal activity
- they have some surfactant properties
- they are readily soluble in water
- they are non-corrosive, and may be left on equipment over-night
- they are stable
- they are relatively non-toxic to mammals.

Disadvantages include being inactivated by dirt, and by anionic surfactants.

The **amphoteric** group of sanitisers ("amphoterics") are very similar to the quaternary ammonium compounds. Their basic structure is:

$$R.NHCH_2.COOH \quad e.g. \quad C_{12}H_{23}.N^+H_2.CH_2COO^-$$

Because of their anionic group, they combine detergency with sanitising activity, and they are alleged to be inactivated by dirt less readily than are the other sanitisers.

6.3.3.4 Phenolic compounds

This group includes a wide range of compounds, loosely based on phenol. They are excellent sanitisers, but suffer from the major disadvantage of having a strong phenolic odour. This precludes their use in food processing areas, or where packaging material is stored, or wherever their use may cause odour to drift into such areas. For this reason their use is limited to toilet areas and outdoor areas, e.g. around drains.

6.3.4   Choice of Sanitiser

Table 6.3 shows a comparison of the three major groups of sanitisers that can be used in food processing areas. The amphoteric sanitisers have similar properties to the quaternary ammonium compounds.

Table 6.3 Comparison of sanitisers.

Sanitiser	Advantages	Disadvantages
Chlorine compounds	Very effective Low cost	Corrosive Can be unstable Potentially dangerous to personnel Activity reduced by dirt Odour

Iodophors	Effective
	Some resistance to dirt
	Colour an indicator of
	  potency
	Non-corrosive (if used
	  properly)
		Slight odour
		May stain equipment

Quaternary Ammonium compounds	Effective Very stable Non-corrosive Non-toxic to mammals	Slight odour Activity reduced by dirt

All of these sanitisers are effective. Some authorities suggest that if an alkaline detergent has been used to remove dirt, a chlorine compound or quaternary ammonium compound should be used as the sanitiser; and if an acid detergent has been applied, then an iodophor should be used. But in practice, provided that the detergent is thoroughly rinsed away before the sanitiser is applied, any type can be used.

The contact time is an important consideration. If using a chlorine compound, it must be thoroughly rinsed away after about 10 minutes, or there could be a corrosion problem. Quaternary ammonium compounds, however, may be left on equipment overnight, to give residual disinfection, and then rinsed off before work starts in the morning.

For selection of a particular brand of a sanitiser type, several laboratory tests are available (Appendix 2). However, for most practical purposes, comparisons should only be made under realistic working conditions. Whichever is chosen, it is important to **always follow the manufacturer's instructions**. Use at too low a concentration will be ineffective. Use at too high a concentration will be a waste of money, and may lead to corrosion problems.

6.4 USE OF HOT WATER TO KILL MICROORGANISMS ON EQUIPMENT

In some situations, e.g. enclosed systems and pipelines, hot water (80°C) or steam may be used rather than a chemical sanitiser for killing bacteria. However, in non-enclosed systems, e.g. a general processing department, hot water is **not** recommended for this purpose. There are several reasons for this:

- water or steam from a hose rapidly loses heat on dispersion to the atmosphere, and so the required heat does not actually reach the equipment
- metal is an excellent heat conductor so hot water will cool rapidly on the equipment. Hence, it is difficult to achieve the required temperature
- excessive steam leads to high humidity, poor visibility and condensation problems
- high temperatures can be dangerous to personnel
- heat can distort equipment.

For a non-enclosed system, hot water is much less satisfactory than a chemical sanitiser for reducing bacterial numbers.

Chapter 7

CLEANING PROCEDURES

7.1 THE CLEANING PROCESS

Cleaning procedures can vary depending on the exact circumstances. However, the following is a basic procedure which can be adapted to meet virtually all situations. The four objectives of cleaning were listed in the previous chapter. These objectives are closely linked in that any process which achieves any one of them is likely to achieve all four. It is clearly established that if all dirt is not removed, the sanitiser will fail. **Thorough cleaning must always precede sanitisation**. In addition to removing dirt, efficient cleaning removes 99% of all microorganisms purely mechanically. However, this cannot always be guaranteed, hence the justification for using sanitisers.

1. **Dismantle** and move equipment where necessary. Clean personal equipment and small items.

2. **Dry clean**. Sweep, scrape, pick up scraps and place in waste areas or bins. Recover as much product as possible for re-working. Do not put solid material into the drains as this will increase the cost of effluent treatment, and feed any rodents that live in the drainage system. Leave grids in place while cleaning. This dry clean is an important step since consider-

able dirt is removed without the use of water or expensive detergent.

3. **Hose with warm water.** This removes much of the excess dirt (without use of a detergent), and "lifts" and "softens" any remaining residues. **Hot** water should not be used as this might cause baking of some protein residues which are subsequently difficult to remove. **Cold** water is less effective at softening dirt.

If these first three steps are performed effectively, most of the gross contamination will be removed, and only stubborn residues will remain.

4. **Apply a suitable detergent** as recommended in the manufacturer's instructions. The detergent will suspend and solubilise the remaining dirt, and the mixture will be hosed to the drains. The amount of manual scrubbing required will depend on the cleaning technique used (refer Section 7.3). If it is necessary to use both an alkaline and acid detergent in the procedure, use the alkaline detergent first, and thoroughly rinse away before application of the acid detergent.

5. **Rinse** with warm or hot water to ensure that all dirt and detergent are washed down the drain. Use a set order of rinsing, starting from the ceiling/tops of walls and working downwards, to ensure that previously rinsed areas are not recontaminated. **Cold** water is not used for this step as it would allow fats to re-solidify. This step is important in that if any dirt or detergent residues remain, the subsequent sanitiser may be inactivated.

6. **Apply sanitiser,** usually by spraying (fogging), according to the manufacturer's instructions. This step ensures the killing of bacteria. Particular attention should be paid to the "contact time", which if insufficient, will give inadequate killing; if too long, there may be corrosion problems, particularly when using hypochlorites.

7. **Rinse with hot water.** This removes all traces of sanitiser which might otherwise contaminate product. A set order of

rinsing should be used, working downwards from the top, to avoid recontaminating areas which have been rinsed previously. It is preferred to use hot rather than cold water for this step since cold water, although cheaper to produce, may contain a relatively high number of bacteria (even if it is potable). Hot water will contain far fewer organisms to contaminate the product-contact surfaces.

8. **Allow to dry.** Do not wipe down with cloths or towels as this will recontaminate the surfaces. Rather, turn on the fans, and position equipment so that it drains dry. Any bacteria which survived the cleaning process will be unable to grow on a dry surface, and may even die. Wet surfaces are more protective of microbial cells.

It is essential that every surface touched by the food product or ingredient be visually clean, free of chemical residues, and not have excessive microbial populations. A commercially-clean surface will not be sterile, but the number of organisms present should be much less than that on the food passing over it. A level of less than $10^2/cm^2$ should be attainable.

7.2 FUMIGATION

Occasionally, it is necessary to fumigate an area, e.g. a chiller. In this situation, a gaseous sanitiser is used as the aim is to kill airborne microorganisms, including many spores, in addition to those on product-contact surfaces. Although it is possible to use an ordinary sanitiser, e.g. a quaternary ammonium compound, or even an acid such as lactic acid, delivered in such a way that droplets are dispersed throughout the air, it is more usual to fumigate using formaldehyde. This is a gas which is very effective against all microorganisms, including spores. It is generated by vaporising paraformaldehyde tablets or by reacting formalin with potassium permanganate. Because formaldehyde is toxic to humans, care must be exercised in its use and rooms must be sealed during fumigation. A typical procedure would be:

1. Defrost where necessary.

2. Remove all product and ingredients.

3. Scrub out with a suitable detergent, to remove all organic material.

4. Steam out. Effective fumigation requires a temperature of at least 18°C, and relative humidity greater than 85%.

5. Vaporise the calculated amount of paraformaldehyde tablets (use the supplier's recommended amounts) by using a "meths" burner, or react the appropriate amounts of formalin and potassium permanganate by dropping the formalin from a separating funnel onto crystals of permanganate.

6. Seal the room, while leaving fans running slowly. Leave for 24 hours.

7. Open the room, and disperse the fumes. Do not enter the room until the fumes no longer irritate the eyes.

Depending on the situation, fumigation, as a preventive measure, is usually performed no more than once or twice a year. Occasionally, however, it may need to be performed in response to defects being detected in the product, or if there is clear visual evidence of, e.g. mould growth on ceilings, necessitating immediate remedial action.

7.3   METHODS OF DETERGENT APPLICATION

There is no single method that can be recommended for all situations. The important considerations are the total cost of the cleaning operation, and the degree of control which is exercised over the process. Labour can account for up to 70% of total cleaning costs, so it may be necessary to use an application technique that minimises this aspect. However, it is important that there is good supervision of cleaning, and that there is full control over chemical usage and the cleaning technique. The following describes the application methods which are available, but the choice is always location-specific.

7.3.1 For "Open" (Non-enclosed) Areas

7.3.3.1 Scrubbing brush and bucket

This is a traditional technique in which the personnel are provided with a bucket containing detergent solution, and a scrubbing brush. Each person is allocated a specific area to clean, and has been given precise instructions on how to do it, e.g. the order in which items are to be cleaned. In small areas, where there is good supervision, this is an excellent technique. However, in large areas, it suffers from the disadvantages of being wasteful of detergent, labour intensive, and difficult to control.

After use, brushes should be thoroughly cleaned and sanitised.

7.3.3.2 High pressure sprays

These were developed in order to reduce the labour costs of cleaning, and to allow more effective control over the operation. High pressure sprays use a hose (Figure 7.1a) to deliver acid or

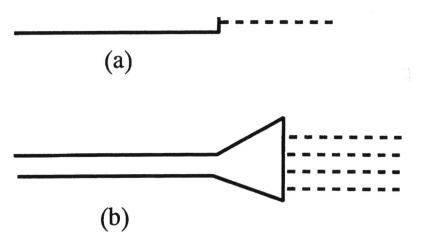

Figure 7.1 (a) Narrow nozzle for use in high pressure sprays. (b) Wide angle nozzle for use in low pressure sprays.

alkaline detergent, at the appropriate temperature, onto the dirty surfaces. Up to 15 litres per minute can be delivered at pressures up to 5000 kPa. The jet of liquid hitting the surface provides mechanical energy to remove the dirt. However, manual scrubbing may still be required in particularly dirty areas.

Although this technique does have the advantages of lower labour costs, and more effective control, it does have some disadvantages. Detergent usage is high, and much of the liquid runs to waste before having a sufficient contact time with the surface. Often, dirt is spread around, even onto previously-cleaned surfaces, due to the high pressure jet of liquid. In addition, high pressure sprays can be dangerous if staff become involved in occasional "horseplay".

These systems can be used as either portable pumps, or *via* centralised pumping stations.

7.3.1.3 Low pressure sprays

These have the same advantages as high pressure sprays, and, in addition, because they provide less mechanical energy, less detergent rebounds off the surface to go to waste or splash previously-cleaned surfaces. By using a wide angle nozzle (Figure 7.1b) a good coverage of detergent can be achieved, which can soak into the dirt and loosen it. A high pressure spray of water can then be used during the rinsing step to remove the dirt/detergent mixture.

Manual scrubbing may still be required for stubborn dirt.

7.3.1.4 Foam cleaning

This is the application of a detergent in such a way that a "foam" is produced. This adheres to the dirty surface (including vertical surfaces) and allows time for the detergent to penetrate and solubilise the dirt. The foam acts by slowly collapsing, constantly bringing layers of fresh detergent into contact with the dirt. The importance of contact time cannot be overstressed; the foam requires time to collapse and thus bring the detergent into contact with the dirt. Manual scrubbing may be required for particularly dirty areas.

Foam is prepared and applied *via* a "foaming unit". This may be a portable system, or using a fixed ring main. Air injection is required to produce the foam.

Advantages of foam cleaning include:

- reduced labour costs
- foam clings to a surface, and does not immediately run to waste
- the operator can easily see where foam has been applied, so coverage is more thorough.

A disadvantage of this technique is its extra complexity, e.g. the rate of air injection must be carefully controlled. Hence, operators and supervisors require additional education and training in its use.

7.3.2   For Enclosed Areas

For these areas (e.g. pipelines, tanks), CIP (cleaning in place) systems should be used.

7.3.2.1 CIP cleaning

This system is entirely automatic and involves the circulation of detergent solutions around the vessels, pipelines, etc. (Figure 7.2). No manual cleaning is required, hence labour costs are minimal. Another advantage is that solutions can be re-used rather than going to waste.

The entry of detergent solutions into large vessels is *via* spray-balls or rotating jets. These ensure that the liquid flow is directed to the sides of the vessel, from where it runs down to be recirculated.

A basic CIP system will include:

- a detergent solution reservoir
- a detergent dosage system
- a temperature control mechanism

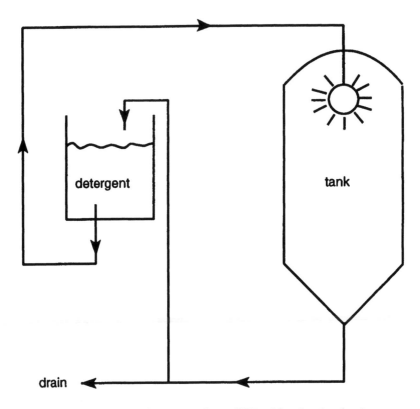

Figure 7.2 Schematic diagram of a C.I.P. (cleaning-in-place) system. Detergent solution is pumped around the vessels and pipelines, and may be recovered or sent to waste, as required.

- a pump, to generate the required turbulence
- a piping system incorporating:
  - some protection mechanism (e.g. micro-switches on valves) to ensure that the CIP fluid is not pumped to the wrong place, or to the right place at the wrong time
  - sprayballs or rotating jets to distribute the fluid over the surface of the vessel.

Normally, there are five successive steps in CIP cleaning:

1. Preparatory steps
   - remove and displace from the system as much product as possible. Discard or reclaim as appropriate;
   - remove, and clean separately, those items of equipment which cannot be effectively cleaned by CIP, or which restrict the flow of the detergent solution, e.g. flow controllers, filtration systems;
   - install "jumper" pipes to isolate the CIP system from the normal production lines while these are still in use during cleaning operations.

2. Pre-rinsing
   - pre-rinse using sufficient quantities of water at temperatures ranging from 10 to 50°C to remove product residues. Adequate rinsing is usually indicated by clean water flowing from the discharge point;
   - pre-rinse solutions should not be recirculated but run to waste;
   - remove, clean and replace those items of equipment which are to be manually cleaned and are necessary for the completion of the circulation CIP circuit, e.g. valves, thermometers.

3. The detergent cleaning cycle
   - the detergent cleaning cycle is performed by circulating an acidic or alkaline detergent solution, as prescribed for the specific item or equipment, at the recommended concentration. The temperatures required depend upon the particular circumstance. Where direct steam is used to raise the temperature of the detergent solution, ensure that the concentration of the detergent solution is not reduced below the stated requirements. Indirect heating of detergent solutions is more satisfactory, and should be used wherever possible;
   - for some items of equipment the detergent cleaning cycle consists of a single detergent treatment requiring the

circulation of either an acidic or alkaline detergent through the system. For other items, the detergent cleaning cycle consists of a sequence of detergent treatments in which each detergent solution is separately circulated through the system. In such sequential detergent cleaning operations, the system is given an intermediate water rinse between the change-over of detergent solutions to remove traces of the detergent previously used. This rinse is carried out at the same temperature as used for the detergent cleaning cycle;

- the detergent solution should be circulated at a velocity and temperature, and for the time required, to allow the detergent to clean the equipment.

4. The post-detergent rinse

- rinse the plant using sufficient quantities of water to remove detergent residues, running the solution to waste. This solution is not recirculated, but, if desired, may be reclaimed and used for subsequent pre-rinsing operations;
- post-detergent rinse solutions are normally of cold water;
- continue flushing until the plant is cool;
- replace those items of equipment that were previously removed and separately hand-cleaned.

5. Sanitising

- sanitise the cleaned equipment within 30 minutes of re-use using **one** of the following techniques:
  - i   circulate potable water, at a minimum temperature of 82°C at the discharge end, through the circuit for a minimum of 5 minutes;
  - ii   apply an aqueous solution of a suitable chemical sanitiser of acceptable strength and at the recommended temperature for a minimum of 2 minutes. The solution must contact all product-contact surfaces;

iii expose the circuit to steam to achieve a minimum condensate temperature of 77°C at the drainage outlet for 15 minutes, or a minimum temperature of 92°C for 5 minutes.

Note that this basic procedure will need to be adapted for particular situations.

7.4 CLEANING TEAMS

For cleaning "open" areas, some manual labour is always involved. Three different types of teams can be employed.

7.4.1 Production Staff

This type of team consists of staff whose full-time job is in production, but who are involved in the cleaning operation at the end of the day's work.

The disadvantage of this system is that the cleaning is done by staff who are tired or who may be anxious to leave the factory as soon as possible. Hence the cleaning may not be performed as well as it should be. In some factories, only selected staff work the extra time doing the cleaning, and receive extra payment. Failure to perform the task properly leads to dismissal from the cleaning team, and replacement by a colleague.

An advantage of this type of team is that the production staff become intimately involved with cleaning, and it becomes part of their normal routine throughout the day. This leads to increased awareness of cleaning and sanitation.

7.4.2 Separate Team

This type of team is employed by the company, but their sole job is cleaning. This team commences work after production has ceased and the staff clean all the areas in the factory.

The advantages of this system are that the staff involved can be thoroughly trained for the task, and it is their main occupation rather than it being secondary to production. In addition, the cleaning is done by fresh, rather than tired, staff. The disadvantage is that the production staff lose some of their involvement in the cleaning process.

7.4.3   Contract Team

This type of team is not employed by the company. In this system, all cleaning operations are contracted out to a company which specialises in factory cleaning. The team arrives at the factory after production has ceased, and cleans all the appropriate areas.

The advantage of this system derives from the use of specialists, but the disadvantage is the lack of involvement and awareness of the production staff.

The type of cleaning team selected will depend on the individual factory circumstances. However, if it is accepted that cleaning and sanitation are the responsibility of everybody, then every employee of the company should be involved. Hence, all production staff should view cleaning as part of their individual jobs.

7.5   EVALUATION OF CLEANING PROCEDURES

Since cleaning is a costly operation, some kind of evaluation is essential. This is normally done during pre-operative inspection (refer Chapter 10) using the following criteria ("touch, sight and smell" tests):

- general appearance; no contamination should be visible under good lighting conditions
- work surfaces should not feel greasy when rubbed with the fingers
- a clean, white tissue should not be discoloured when rubbed over the surface of cleaned stainless steel (does not apply to aluminium or galvanised material)
- objectionable smells should not be noticeable

- cleaned surfaces should not show signs of excessive "water break" when wetted.

In addition, a **bacteriological assessment** should be made of the cleaned surfaces (Appendix 4). Because the results of bacteriological tests come too late for any immediate remedial action, their purpose is to:

- **pinpoint trouble spots.** Some items of equipment may regularly give high bacterial counts. These items can be identified to the cleaners so that they are given particular attention;
- **monitor trends.** Normally, the bacterial count should be less than $10^2/cm^2$. Most counts will be well below this value, and the occasional higher count is not a significant problem provided it is not always from the same item of equipment. However, if the **average** count is determined, on a weekly basis, then it is possible to follow the general trend in results. If this trend is upwards, this signifies a deterioration in the standard of cleaning, and steps should be taken to find the cause and remedy the situation (Figure 7.3).

More detail on the value of bacterial counts in monitoring the effectiveness of cleaning procedures is given in Chapter 11.

7.6 TROUBLESHOOTING OF CLEANING PROCEDURES

Sometimes, cleaning procedures fail, e.g. dirt is not removed or bacteriological counts are high. If this happens, it is important to find the exact cause of the problem rather than to simply increase the strength of the detergent solution, or use it at a higher temperature. Such remedies may not solve the problem, and may cause others, e.g. corrosion of metals. The main reasons for failure can be listed:

- wrong detergent for the job. The type of detergent being used should be reconsidered;

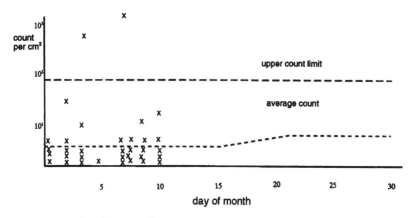

Figure 7.3 Plot of bacterial counts on equipment over a period of time. The bacterial counts that are obtained each day are plotted on the graph so that troublespots can be pinpointed and trends can be monitored. If the same item of equipment regularly gives counts above the "upper count limit", particular attention should be paid to this item during cleaning. If the average count trends upward, this signifies that there is a deterioration in the standard of cleaning, and action should be taken to find the cause and remedy the situation.

- the detergent is not being used according to the manufacturer's instructions. Particular attention should be paid to concentration and temperature of use, and contact time;

- poor planning of cleaning procedure. For example, the order of cleaning particular items may be poor, so that cleaned areas become recontaminated during rinsing. Particularly dirty areas may need cleaning more than once during a day. Insufficient scrubbing may leave stubborn deposits;

- poor control and supervision of cleaning procedure, i.e. the planning is sound but the execution is uncontrolled. Assuming that the supervisor has been trained in sanitation procedures, this problem is one of management rather than of poor planning;

- sanitiser is not being used according to the manufacturer's instructions;

- microorganisms are becoming resistant to the sanitiser. This sometimes occurs when using, for example, quaternary ammonium compounds on a regular basis. The remedy is to occasionally use a different type of sanitiser, e.g. a hypochlorite.

Chapter 8

WATER TREATMENT

All cleaning procedures use water for rinsing purposes, and water is an ingredient in many foods. Obviously, such water should not be contaminated with microorganisms, or else product-contact surfaces and food could become contaminated. All water used in food processing areas, either as an ingredient or for cleaning purposes, should be **potable**, i.e. of a standard high enough for human consumption.

The water used in food processing originates in rain. Rain is not sterile; it collects dust particles and microorganisms on its way to the ground. Once on the ground, more organisms are collected from the soil, e.g. *Bacillus, Clostridium, Pseudomonas*. Thus, the water entering streams and rivers contains a range of microorganisms, but their numbers are relatively few due to lack of nutrients in the water.

In many cases, rainwater does not collect in streams and rivers, but percolates down through the soil and, eventually, through the rock beneath the soil. Often, the rock structure filters out the microorganisms. **Boreholes** and **artesian wells** utilise this water which is virtually uncontaminated. However, in many areas, this water may be contaminated with chemicals which have originated from industry or agriculture, e.g. nitrates, and which have found their way into the groundwater.

Although the number of microorganisms in streams and rivers is relatively few, due to lack of nutrients, this situation can change drastically if the nutrient content is increased. This can occur following run-off of faeces or fertiliser from the land, or from the input of domestic or industrial sewage into the river. In this case, pathogens of faecal origin may also be present. For these reasons, river water usually requires treatment before it is used in food processing areas.

The amount of water treatment required depends on the quality of the raw water, e.g. water from a lowland river may require a whole range of treatment processes, whereas that from an artesian well may require only minimal treatment. This chapter outlines a whole range of treatment methods which may be applied to the production of potable water.

8.1   PURPOSE OF WATER TREATMENT

The purpose of water treatment is to reduce to acceptable levels, as laid down in legislation, the amounts of:

- organic chemicals
- inorganic chemicals
- colour
- turbidity (suspended or colloidal solids)
- microorganisms.

8.2   PRE-TREATMENT METHODS

Depending on the circumstances, some pre-treatment may be applied prior to the main treatment facility.

8.2.1   Screening

To ensure efficient and reliable operation of the main units in a treatment plant, it is often necessary to first remove large floating objects, e.g. branches in river water, which could obstruct flow in the plant. For most waters, screens of 5–20 mm aperture mesh are used for this purpose. Such screens are usually cleaned automatically, using rakes, to prevent blockage.

The **microstrainer** is a development of the drum screen and uses a fine, woven stainless steel mesh, supported on a coarser mesh to give strength (Figure 8.1). The straining mesh is available in various grades from 26 to 65 µm. Typically, the microstrainer is 4 m in diameter and 4 m in length, and rotates at a maximum peripheral speed of 0.5 m per second. Hydraulic loadings are in the range 700–2300 m^3 of water per m^2 of mesh per day. Because of the small apertures, clogging can occur rapidly and so the mesh is continually washed by pressure sprays. These washing jets use about 2% of the total output of the microstrainer.

The microstrainer can remove large amounts of suspended solids from water, including algae, but not bacteria. If the raw water is clean enough, e.g. from an upland river, this equipment alone may provide sufficient treatment, combined with chlorination.

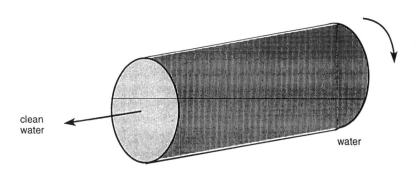

clean water

water

Figure 8.1 The microstrainer, as used in water treatment.

8.2.2 Storage

Self-purification of water during storage in **balancing reservoirs** will lead to a considerable improvement in water quality, particularly with raw waters containing high levels of suspended solids. Depending on the climate, a 10–15 day detention time is probably sufficient to remove most suspended material and bacteria, without leading to development of troublesome algal growths.

8.2.3 Aeration

Introduction of oxygen into the water helps to remove dissolved gases such as carbon dioxide and hydrogen sulphide. Thus, the odour and taste are improved. In addition, if the water contains high concentrations of iron or manganese salts, these will be oxidised and removed as settleable solids. Without prior removal, iron and manganese salts could cause subsequent problems by being deposited in pipelines and causing blockages.

Methods of aeration include spillage of the water over weirs, the use of spray nozzles, and direct injection of compressed air.

8.3 TREATMENT METHODS

8.3.1 Chemical Coagulation

The purpose of this technique is to remove suspended solids from the water. In theory, if water is held still (quiescent) for a sufficiently long period of time, all suspended solids will settle out by sedimentation. Unfortunately, in practice, this is difficult to achieve due to the difficulties of keeping water quiescent, and because small particles and colloidal material (which often cause turbidity and color in water) have extremely low settling rates and are not easily removed.

Chemical coagulation involves the addition of a chemical to the water which accelerates the removal of suspended solids. **Alum** (aluminium sulfate) is the most widely used coagulant.

When added to water, alum reacts to form aluminium hydroxide, which exists as a "floc" precipitate. This precipitate carries a nett positive charge, and so it acts as a magnet, drawing to itself negatively charged suspended and colloidal solids. Because the "floc" precipitate settles out relatively rapidly, the entire sedimentation process is very much accelerated. Typical equipment used for chemical coagulation is shown in Figure 8.2. The coagulant is added to the water prior to entry into the draught tube. The "floc" precipitate which is formed, together with the attached suspended solids, is collected from the bottom of the tank as sludge, while treated water is collected from the top.

Occasionally, polyelectrolytes are used as the coagulant, rather than alum.

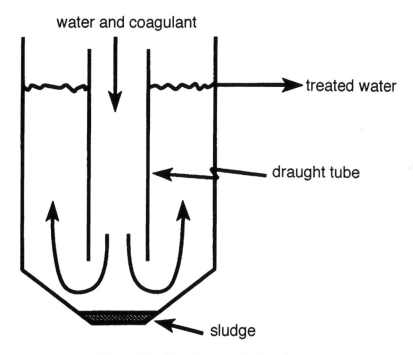

Figure 8.2   Chemical coagulation of water.

8.3.2   Rapid (Pressure) Filtration

Filtration of water, containing suspended solids, through porous media, e.g. sand, is an important stage in the treatment of water to achieve final clarification. Although about 90% of the turbidity and colour is removed by chemical coagulation, a certain amount of "floc" may be carried over and requires removal. Hence, chemical coagulation is usually followed by filtration.

The medium used in a rapid filter is usually sand, with an average particle diameter of 0.5–1.0 mm, packed in a bed of 1–2 m depth. Water, which contains some "floc" material, is passed through the bed of sand, and the "floc" is filtered out by the sand particles. In turn, the "floc" acts to filter out any other suspended or colloidal material from the water. When operating under gravity conditions, hydraulic loadings of 120 m^3 of water per square metre per day can be achieved. Higher loadings can be achieved when the filter is operated under pressure. To remove the "floc" particles which accumulate in the sand, the filter is cleaned every 1–3 days by backwashing with clean water. Some-times, the backwashing is preceded by agitation of the sand with compressed air.

The normal method of filtration downwards through a sand medium is sometimes inefficient because the main solids load falls on the particles at the top. This causes blockage, and the solids must be removed by backwashing. It would be more effi-cient to have larger particles, up to 3 mm diameter, at the top to remove large solids, thus reserving smaller sand particles to trap the really fine solids. This can be achieved using an upper bed of anthracite (particle diameter 1.25–2.50 mm) on top of the lower sand bed. Alternatively, upward filtration can be used whereby an upper sand layer is placed above a gravel layer (particle diameter 1–2 mm), and the water is pumped upwards through the filter. In this way, the suspended solids in the water meet the large gravel particles first. Care must be taken to control the filtration rate of this method since high velocities will expand the bed and allow solids to escape.

8.3.3 Carbon Filtration

Following sand filtration, the clarified water may be further treated by passage through an **activated carbon filter**. This removes organic materials from solution, including those causing color, taste and odour, by adsorption onto the carbon.

Activated carbon filters can also be used to remove chlorine from water, particularly when the water is to be used as an ingredient, e.g. in the brewing industry.

8.3.4 Water Softening

Hardness in water may be either temporary or permanent, and is caused by the presence of calcium and/or magnesium salts in the water.

Temporary hardness	Permanent hardness
calcium bicarbonate	calcium chloride
magnesium bicarbonate	calcium sulfate
	magnesium chloride
	magnesium sulfate

A soft water will have less than 60 mg/l of hardness (expressed as calcium carbonate), whereas a hard water will have up to 300 mg/l of hardness.

Temporary hardness will be deposited as calcium or magnesium carbonate when water boils. In a food processing factory this will lead to reduced heat transfer in the boilers,

$$\text{e.g.} \quad Ca(HCO_3)_2 \xrightarrow{\text{heat}} CaCO_3 \downarrow + CO_2 + H_2O.$$

Permanent hardness can react with some detergent ingredients, e.g. sodium hydroxide, to form a deposit of calcium (mineral) scale on equipment.

$$\text{e.g.} \ 2\,NaOH + CaCl_2 \longrightarrow Ca(OH)_2 \downarrow + 2\,NaCl.$$

For these reasons, water for use in a food processing factory may require softening. The following two techniques are available.

8.3.4.1 Chemical precipitation

In this technique, chemicals are added to the water to react with the hardness, and to remove the calcium and magnesium salts as precipitates.

The addition of **lime** ($Ca(OH)_2$) to water removes temporary hardness by precipitation of calcium carbonate and magnesium hydroxide:

$$Ca(HCO_3)_2 + Ca(OH)_2 \longrightarrow 2CaCO_3 \downarrow + 2H_2O$$

$$Mg\,(HCO_3)_2 + 2Ca(OH)_2 \longrightarrow 2CaCO_3 \downarrow + Mg(OH)_2 \downarrow + 2H_2O$$

The effect of lime on magnesium permanent hardness is to convert it to calcium permanent hardness:

$$MgCl_2 + Ca(OH)_2 \longrightarrow Mg(OH)_2 \downarrow + CaCl_2$$

$$MgSO_4 + Ca(OH)_2 \longrightarrow Mg(OH)_2 \downarrow + CaSO_4$$

Lime has no effect on calcium permanent hardness, hence the nett result of lime addition is the removal of temporary hardness and of magnesium permanent hardness, leaving only calcium permanent hardness.

The addition of **soda** (sodium carbonate, Na_2CO_3) to water removes calcium permanent hardness by precipitation as calcium carbonate:

$$CaSO_4 + Na_2CO_3 \longrightarrow CaCO_3 \downarrow + Na_2SO_4$$

$$CaCl_2 + Na_2CO_3 \longrightarrow CaCO_3 \downarrow + 2NaCl$$

In practice, a combined **lime-soda softening** process is used. The lime and soda are often added to the water in conjunction with alum, and the precipitates are removed in the chemical coagulation tank and during subsequent filtration. In this way, the water is adequately softened.

8.3.4.2 Ion-exchange process

In this technique, the water is passed through a cation exchange resin, which is usually in the sodium form. The sodium ions on the resin exchange with the calcium and magnesium ions in the water, thus softening the water (Figure 8.3). In contrast to the lime-soda process, this method does not decrease the total dissolved solids content of the water. After all of the sodium ions on the resin have been exchanged, it is necessary to regenerate the resin. This is achieved by passage of a strong brine (NaCl) solution through the resin so that the calcium and magnesium ions are forced off and into the water, leaving the resin in the sodium form. The small amount of water containing the calcium and magnesium ions should not be used in the factory.

It is possible to use a combination of ion exchange resins, so that completely demineralized water is produced. In the first resin, which is a cation exchanger in the hydrogen ion form, calcium and magnesium ions are exchanged for hydrogen ions:

$$\text{e.g. } CaCl_2 + 2\text{H-resin} \longrightarrow 2HCl + \text{Ca-resin}$$

The hydrochloric acid which is produced causes the water to become acidic. In the second resin, which is an anion exchanger in the hydroxyl ion form, chloride ions are exchanged for hydroxyl ions:

$$\text{e.g. } HCl + \text{HO-resin} \longrightarrow H_2O + \text{Cl-resin}$$

Thus, the acidity is removed in addition to the water being completely demineralized. For regeneration, hydrochloric acid is passed through the cation exchanger, and sodium hydroxide through the anion exchanger.

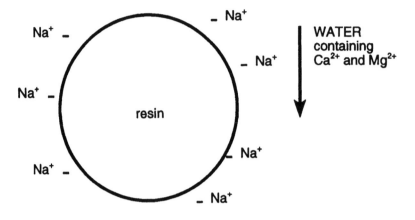

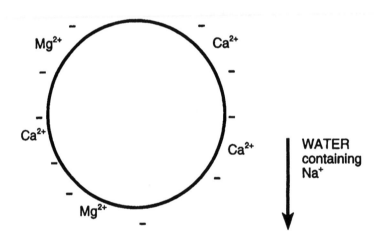

Figure 8.3   Ion exchange process for softening water. During passage through the resin, the calcium and magnesium ions in the water exchange for sodium ions on the resin.

8.3.5   Chlorination of Water

The primary purpose of chlorination is the control of microorganisms in the water, i.e. it is the addition of chlorine to the water to kill pathogens. Chlorinated water, however, is not necessarily sterile, even if it meets the legislative requirements for potable water.

A secondary purpose of chlorination is the removal, by oxidation, of certain chemical contaminants, e.g. iron and manganese salts, hydrogen sulphide.

To understand the practical aspects of chlorination, it is necessary, first, to understand the chemistry of the process. When chlorine is added to **pure** water, the following reaction occurs:

$$Cl_2 + H_2O \longrightarrow HCl + HClO \qquad (1)$$

The effective sanitising agent is hypochlorous acid (HClO). The chlorite ion (ClO^-) is relatively ineffective at killing bacteria. Since the dissociation of hypochlorous acid is suppressed at acid pH values, it follows that killing is more effective in acid conditions. The concentration of hypochlorous acid is referred to as the **free chlorine residual** (free available chlorine, FAC).

If any oxidizable organic or inorganic chemicals are present in the water, chlorine will react with them, and will, thus, be unavailable to form a free chlorine residual, i.e. the chlorine is inactivated;

e.g. chlorine is inactivated by hydrogen sulphide

$$4Cl_2 + H_2S + 4H_2O \longrightarrow H_2SO_4 + 8HCl \qquad (2)$$

This reaction is referred to as the **chlorine demand** of the water, and is a measure of the concentration of contaminating substances. A water with a high chlorine demand will inactivate more chlorine than that with a low chlorine demand.

When ammonia is present in the water, the following series of reactions occurs:

$$
\left.\begin{array}{ll}
Cl_2 + NH_3 \longrightarrow NH_2Cl + HCl \\
Cl_2 + NH_2Cl \longrightarrow NHCl_2 + HCl \\
Cl_2 + NHCl_2 \longrightarrow NCl_3 \quad + HCl
\end{array}\right\} \quad (3)
$$

The monochloramine, dichloramine and nitrogen trichloride which are formed are known collectively as a **combined chlorine residual**. Organic chloramine compounds can also be formed. These combined residuals have disinfectant activity, but they are less effective than the free residual in that 100 times the contact time is required to achieve a similar kill.

Once all the ammonia has reacted, further chlorine converts the combined residual into a free residual:

$$ NCl_3 + Cl_2 + H_2O \longrightarrow HClO + NH_4{}^+ \qquad (4) $$

If any oxidizable organic or inorganic chemicals remain in the water, hypochlorous acid will react with them, and will be inactivated.

e.g. hypochlorous acid will be inactivated by nitrite ions:

$$ NO_2{}^- + HOCl \longrightarrow NO_3{}^- + HCl \qquad (5) $$

For most waters undergoing chlorination, the sequence of the above reactions is (2), (3), (4), (5), (1). Only after all oxidizable material in the water has been oxidized does a free chlorine residual appear. This point is known as the **breakpoint**, after which the free chlorine residual produced is directly proportional to the chlorine dosage. Figure 8.4 illustrates the situation by showing a plot of the chlorine residual against the chlorine dosage into the water.

A major point arising from this graph is that, because of the chlorine demand of the water, the free chlorine residual is always less than the chlorine dosage, e.g. to achieve a free chlorine residual of 0.3 ppm, it may be necessary to dose 1.0 ppm chlorine. Because the chlorine demand and ammonia content of the water

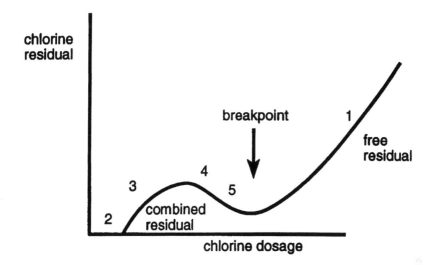

Figure 8.4 A plot of chlorine residual against chlorine dosage during chlorination of water. The numbers refer to the reactions described in the text: 1. Formation of a free chlorine residual from chlorine. 2. Chlorine demand (inactivation of chlorine by oxidizable material). 3. Formation of a combined chlorine residual. 4. Formation of a free chlorine residual from a combined chlorine residual. 5. Chlorine demand (inactivation of free chlorine residual by oxidizable material).

can change daily, it is often necessary to adjust the chlorine dosage daily to attain the same free residual.

Three types of chlorination method may be practised:

- **marginal chlorination**. This represents the addition of sufficient chlorine to the water to achieve a combined chlorine residual. It should be used only for waters which are virtually uncontaminated by bacteria. It is not recommended for the food industry, and, in most countries, legislation prevents its use;

- **breakpoint chlorination**. This represents the addition of sufficient chlorine to achieve a free chlorine residual. The exact concentration is normally determined by legislation, but a typical value of the free chlorine residual would be 0.3 ppm after 20 minutes contact time, or at the point of usage. Although such water may be potable according to the regulations, it must be emphasised that potable water is not necessarily sterile water. Although most pathogens and coliform bacteria are killed by a free chlorine residual of 0.3 ppm, many other bacteria will survive, and the water can serve as a source of contamination for food product. The simple remedy of using a higher value of free chlorine residual is not always practical, since it will involve higher costs, may cause the water to smell and/or taste of chlorine, and may lead to corrosion problems. If there is a problem of contaminating bacteria in potable water, a more useful remedy is to investigate/ improve the water treatment methods prior to chlorination. The simple remedy of overdosing the water with chlorine, followed by its removal is described next;

- **superchlorination**. This represents the addition of sufficient chlorine to achieve a free chlorine residual of up to 10 ppm. The technique is usually used only in emergency situations since the water will taste and smell strongly of chlorine. The excess chlorine can be removed by treatment with sulphur dioxide or by passage through an activated carbon filter.

8.3.5.1 Methods of chlorine application

Chlorine may be added to water in the form of chlorine gas, sodium hypochlorite or calcium hypochlorite. Where large volumes of water are to be chlorinated, chlorine gas is preferred because:

- it has a total available chlorine content of 100%
- it is the cheapest source of chlorine on the basis of unit available chlorine content
- it is easy to control and apply
- it lowers the pH slightly, thus aiding disinfection.

Certain precautions should be taken when using chlorine gas:

- it is toxic to humans. Being 2.5 times heavier than air, all installation should be at ground level and clear of all basement areas;

- all cylinders should be positively restrained for safety by means of chains firmly attached to the wall or other suitable point;

- the flow rate of gaseous chlorine is temperature-dependent. Hence, the bottle room should be maintained at a constant temperature, e.g. 20°C.

Sodium and calcium hypochlorites require mixing with water in suitable containers (non-metallic) to achieve a solution suitable for dosing. They can then be pumped using relatively simple units.

Regardless of the source of chlorine, only automatic equipment which doses the chlorine in proportion to the **water flow rate** should be used. This is necessary to avoid fluctuations in the chlorine residual which, if too low, would be ineffective, or, if too high, could lead to a chlorine taste and corrosion problems. **Chlorine residuals** in the water should be checked on a daily basis, and the dosage adjusted appropriately.

8.4   STANDARDS FOR POTABLE WATER

Standards vary from country to country, and readers are advised to contact their own local regulatory agency regarding the standards required in their particular factory. The following lists those items which are normally included in the regulations, with typical guideline values, based on World Health Organization publications.

8.4.1   Inorganic Chemical Constituents of Health Significance

Inorganic chemical	Concentration (g/m^3)	
arsenic	0.05	
boron	0.5	
cadmium	0.005	
chromium	0.05	
cyanide	0.1	
fluoride	1.0	(deliberately added)
lead	0.05	
mercury	0.001	
nitrate	10	
selenium	0.01	

8.4.2   Organic Chemical Constituents of Health Significance

Organic constituents	Concentration (mg/m^3)
aldrin and dieldrin	0.3
benzene	10
benzo[a]pyryene	0.01
chlordane	0.3
chloroform	30
2,4,5-T	10
2,4-D	100
DDT and breakdown products	1
diquat	60
gama-HCH (lindane)	3
paraquat	10
pentachlorophenol	10
tetrachloroethene	10
trichloroethene	30
2,4,6-trichlorophenol	10

8.4.3 **Aesthetic Quality.** While the following concentrations are not known to be harmful to health they have been shown to cause the following undesirable effects.

Substance	Concentration (g/m^3)	Effect
aluminium	0.05	discoloration, corrosion
chloride	100	corrosion
copper	0.05	taste, discoloration, corrosion
hydrogen sulfide	not detectable by consumer	taste, odour
iron	0.1	taste, discoloration, turbidity, deposits
manganese	0.05	taste, discoloration, turbidity, deposits
sodium	100	taste
solids (total dissolved)	500	taste
sulfate	50	corrosion
zinc	5	taste, deposits, discoloration

Physical property	Concentration	Effect
color	5 color units	discoloration
hardness	80 g/m^3	scale formation
pH range	pH 7.4–8.5	corrosion, scale
turbidity	1 turbidity unit	discoloration

8.4.4   Microbiological Quality

No coliform organisms should be present in 100 ml of water. If the occasional coliform is detected, it must not be a faecal coliform (*Escherichia coli*).

Procedures for sampling of water and conducting coliform counts are described in Appendix 3.

8.5   THE FUTURE OF WATER SUPPLIES

Increasing industrialization and the use of water for irrigation of agricultural lands is placing severe constraints on the future availability of water supplies. Put simply, there will be less cheap water available in the future. The food industry must approach this problem in two ways:

1. The volume of water used during routine factory operations must be minimised. In addition to simple water-saving techniques, e.g. turning off taps when not in use, engineers must look carefully at the processing operations with a view to their re-design in such a way that water use is reduced.

2. There must be a decrease in the amount of water that is discharged as waste, and an increase in the amount that is recycled. Two broad strategies can be adopted to achieve this:
   - relatively clean wastewater, e.g. cooling water, should not be mixed with dirty wastewater. Rather, it should be collected separately and recycled, after minimal treatment, to ensure it is of a potable standard;
   - the effluent treatment system in the factory should aim to treat the dirty wastewater to such an extent that it can be recycled in some way.

These concepts of waste minimisation and water recycling will be addressed in more detail in Chapter 9.

Chapter 9

EFFLUENT TREATMENT

During their processing activities, most food factories produce some waste materials, or effluents, which require disposal. If these effluents are discharged directly into **receiving waters** (e.g. rivers, lakes, estuaries) without any treatment, there may be some adverse effects on the environment.

Due to the organic material in the effluent there will be an increase in the organic material in the receiving water. Bacteria in the receiving water will use this organic material as nutrient, and will rapidly increase in number. As they grow, the bacteria will utilise oxygen which is dissolved in the water (known as "dissolved oxygen"). Although the dissolved oxygen can be replenished by natural means, the bacteria may use it faster than it can be replenished. Hence, the dissolved oxygen levels decrease and eventually become exhausted. As the conditions become anaerobic, plant and animal life in the water die, and anaerobic bacteria are able to grow, producing foul odours.

Particulate material in the effluent will cause turbidity in the receiving water, thus adversely affecting its aesthetic quality. Turbidity may also prevent light from reaching aquatic plants, causing them to die. If the particulate material settles on the bottom of the receiving water, it may prevent oxygen from reaching bottom-living organisms, again causing their death.

Effluents may contain chemicals which are toxic to plant and animal life, and pathogenic bacteria which can be a source of disease to animals or humans.

It is clear, therefore, that discharge of waste materials can adversely affect the aesthetic quality of receiving waters and their recreational use. Perhaps more importantly, in many localities, the receiving water that is used to dispose of the effluent from one factory is subsequently used to provide the potable water of another factory or community. Hence, there is a strong relationship between water treatment and waste treatment. For this reason, the concept behind the waste treatment legislation in many countries is that the treatment should be sufficient to prevent any adverse effects of the effluent on the receiving water. In general terms, this can be viewed as treating the effluent to:

- remove particulate material
- remove organic material
- remove color
- remove toxic chemicals (including pH adjustment)
- remove pathogenic microorganisms.

In most countries, the extent of treatment is determined by the appropriate legislation.

Because of the complexity of most effluents, their "strength" can be expressed in several ways. One of the most important factors that we need to know about a particular effluent is how much **oxygen** will be required for its degradation (**stabilisation**). On this basis, it can often be decided how much treatment is necessary before the effluent can be safely discharged. The following two terms are commonly used for this purpose:

1. **Chemical Oxygen Demand (C.O.D.)** — This is the amount of oxygen required to completely oxidise a waste. The value is obtained chemically, by reacting the waste with a strong oxidising agent such as potassium dichromate.

   The C.O.D. measures the maximum possible oxygen requirement since it includes all oxidisable material, rather than just biodegradable, or organic, material.

A typical C.O.D. value for domestic sewage would be 1000 mg/l. Values for food factory effluents are often much higher.

2. **Biochemical Oxygen Demand (B.O.D.)** — This is defined as the weight of dissolved oxygen required for the biochemical (biological) oxidation of an effluent during five days incubation at 20°C. It is often referred to as $B.O.D._5$.

    Because the B.O.D. is measured biologically, it accounts for only those materials which are biodegradable. Hence, if an effluent contains a lot of material which is chemically oxidisable but not biodegradable, it will have a high C.O.D. but a low B.O.D. In terms of its effect on the dissolved oxygen content of a receiving water, such an effluent would be relatively innocuous. However, the high C.O.D. does signify chemical contamination, possibly with toxic chemicals. If an effluent contains material which is largely biodegradable, then the B.O.D. value will approximate the C.O.D. value, although it can never exceed it. This reflects the situation for the effluent from most food industries. A typical B.O.D. value for domestic sewage is 800mg/l. Most food industry effluents have higher values than this, but much depends on the processing operations in the factory. Overall, the B.O.D. value signifies the biodegradable, organic material content of the effluent.

    Some other terms which are regularly used to quantitate the strength of an effluent include:

    *   **total solids**. This includes the whole amount of pollutional matter, either organic or inorganic, soluble or insoluble;
    *   **dissolved solids**. This includes the material in solution, either organic or inorganic;
    *   **suspended solids**. This comprises those materials that are not dissolved;
    *   **settleable solids**. This comprises those suspended solids which settle to the bottom under specified conditions;

- **total organic carbon.** This includes all organic compounds, either biodegradable or non-biodegradable.

9.1 PRINCIPLES OF EFFLUENT TREATMENT

1. Suspended solids can be removed by allowing them to settle (sediment), thus leaving a clear liquid effluent.
2. Organic material in solution, which contributes the bulk of the B.O.D., can be removed by one of two methods:

9.1.1 Chemical Treatment

This involves the addition of a chemical which causes the organic material to precipitate or flocculate. The precipitate can then be removed by sedimentation.

9.1.2 Biological Treatment

The organic content of the effluent is removed by allowing bacteria to grow in the effluent. This is essentially the same as what would happen if the effluent were discharged directly into a receiving water, but in this case the removal of organic material occurs on the company's premises.

Biological treatment can occur under aerobic or anaerobic conditions.

9.1.2.1 Aerobic

Organic material + bacteria + oxygen $\longrightarrow$
more bacteria + carbon dioxide + water

The B.O.D. (organic material) in solution is converted to bacterial cells and carbon dioxide. The carbon dioxide escapes to the atmosphere, but the new bacterial cells still constitute B.O.D. since they are simply organic material in suspension. However, since they are in suspension, they can be removed by sedimentation, thus reducing the B.O.D. of the liquid effluent.

9.1.2.2 Anaerobic

Organic material + bacteria $\longrightarrow$
more bacteria + carbon dioxide + methane + water

The B.O.D. in solution is converted to bacterial cells, carbon dioxide and methane. The carbon dioxide and methane escape to the atmosphere, while the bacterial cells can be removed by sedimentation.

In comparison with aerobic treatment, anaerobic treatment has the disadvantage of being slower, but the advantage that fewer new bacterial cells are produced. This is important since although the bacteria can easily be removed from the effluent, they still constitute a waste material, and require disposal by some means.

9.2 FOOD INDUSTRY EFFLUENTS

It is impossible to generalise about the composition of waste streams in the food industry. However, due to their biological nature, they are usually readily biodegradable, and often stronger than domestic sewage.

9.3 TREATMENT PROCESSES

It is difficult to describe every treatment method that is available, since the choice of method depends largely on the characteristics of the waste material. Instead, some broad concepts and techniques will be described since they form the basis of most treatment methods.

9.3.1 Grit Removal and Screening

The crude waste stream is first passed through a grit removal tank where grit particles are rapidly removed by settling. The waste stream is then passed through **screens**, where the objective is to remove large pieces of solids. An example of a type of screen widely used for food industry wastes is the **contra shear** (Figure 9.1). Here, the waste stream is passed into the drum, which is rotating slowly. The liquid, and small particulate material, pass through the mesh while the large solids are screened out and pushed to the solids-collection vessel. The liquid stream is then sent for further solids removal, while the large solids are disposed of by some other means (see section 9.3.4).

9.3.2 Solids Removal

This stage of waste treatment is often referred to as **primary treatment**. In its simplest form, solids removal is achieved by passing the effluent into a **sedimentation tank**, where the flow rate of the liquid is slowed sufficiently so that solid particles have time to sediment to the bottom of the tank while fatty material rises to the surface. Mechanical scrapers then remove the sludge and fat from the tank bottom and top, respectively. These solids can be further disposed of or recycled to the factory as appropriate.

In practice, the efficiency of sedimentation tanks in removing fat and suspended solids is very variable, since removal depends on:

- the detention time, i.e. the time allowed for settling. Detention times of 1 hour are often required;

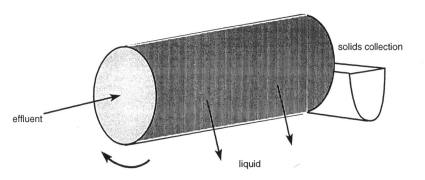

Figure 9.1 A contra shear.

- quiescent conditions, i.e. there should be no disturbances of the effluent from non-baffled inlets or the operation of the scrapers;
- the nature of the waste material being treated.

For these reasons, modern primary treatment methods often involve the use of **Dissolved Air Flotation** (DAF) units (Figure 9.2). Compressed air is injected into the waste stream in an aeration tank, which is enclosed, and the stream is then passed to the flotation tank. As the air comes out of solution it forms bubbles which rise to the surface of the tank. As the bubbles rise they become attached to fat and other solid particles and thus carry them to the surface. From here, the solids can be removed using a scraper mechanism. Any solids which fall to the base of the tank can also be removed using scrapers. This treatment method works particularly well with those effluents that contain considerable quantities of fat.

In the treatment of wastewaters using dissolved air flotation, coagulants are often used to aid the process. Coagulants act by

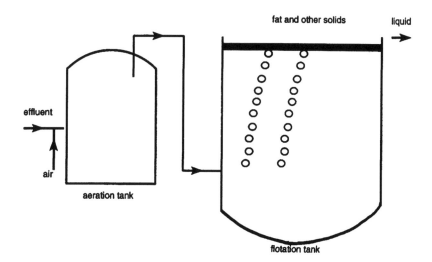

Figure 9.2 A Dissolved Air Flotation unit.

forming stable flocs with the suspended particles in the water. Alum is the coagulant most commonly used, often in association with a cationic or anionic polyelectrolyte, depending on the charge characteristics of the suspended solids. Other inorganic coagulants sometimes used include ferric chloride and calcium hydroxide (lime).

9.3.3   Secondary Treatment

The purpose of secondary treatment is to remove organic material (i.e. soluble B.O.D.) from solution. The two techniques which are available for this are biological secondary treatment and chemical treatment. In addition, "land treatment" may be considered to be a third technique although it is essentially a variant of biological treatment.

9.3.3.1   Biological secondary treatment

In simple terms, this is a relocation to the factory premises of the natural biological cycle, and the process is then performed under controlled conditions. Since the rate of degradation of the organic material is often dependent on the concentration of microorganisms present, it is possible to develop high-rate processes by increasing the microorganism concentration.

Several methods of biological treatment are available. One important factor which governs the choice of method is subsequent solids (sludge) disposal. This is expensive, and it is often economically advantageous to use a method that produces minimal quantities of sludge. However, other factors also need to be considered including the land area available, the location of the premises, the strength of the waste, and the capital and operating costs. If the waste stream is of a particularly high strength it may be necessary to use two treatment methods in series. Also, because of the difficulty of maintaining them aerobic, high-strength wastes often require anaerobic rather than aerobic treatment methods, with the subsequent drawback of odour production. The following describes some of the methods of biological secondary treatment that are available.

Lagoons, or stabilisation basins

These have proved a very popular method of biological treatment of food processing wastes, particularly in rural areas, because of their low capital and running costs, their simplicity of operation, and their ability to produce a reasonable quality, stable effluent. Complete treatment can sometimes be achieved in shallow aerobic lagoons, but a common system consists of a primary anaerobic basin followed by aerated or aerobic lagoons.

Anaerobic lagoons are basins, 3–6 metres deep, in which bacteria break down the wastes to methane, carbon dioxide and bacterial cells in the absence of oxygen. When operating correctly, the gas bubbles which are generated often carry some solids to the surface, forming a crust which prevents the escape of odours and reduces the cooling effect of the outside environment. In other cases, operators may use artificial covers for the same purpose.

The wastewater should enter the anaerobic lagoon near the active sludge layer (i.e. bacteria) on the bottom, and leave at the opposite end near the surface. When operating correctly, there will be little accumulation of additional sludge since it will self-degrade. However, anaerobic lagoons are usually drained annually to allow sludge removal.

Design criteria will depend on the waste being treated, but in temperate climates, a detention time of seven days and a B.O.D. loading of 300 kg per 1000 m^3 per day is normal.

Advantages:	low capital costs; low operating costs; can tolerate high loadings; little sludge produced.
Disadvantages:	large area required, due to long detention times; potential odour problem.

Aerobic lagoons, or oxidation ponds, have a higher surface to volume ratio than anaerobic lagoons. Their depth is usually 0.5–1.5 metres. Often, sunlight is able to penetrate to the bottom of the pond, and green algae use the sunlight, by photosynthesis, to provide a steady supply of oxygen for aerobic bacteria to use during the breakdown of organic material.

Oxidation ponds cannot tolerate high B.O.D. loadings; they would go anaerobic, and thus lose their desired quality of producing no odour. In practice, oxidation ponds are often used after anaerobic lagoons to "sweeten" the effluent prior to discharge into the receiving water.

Advantages:	low capital costs; low operating costs; no odour; little sludge produced.
Disadvantages:	large area required; cannot tolerate high loadings.

Aerated lagoons are characterised by the use of aeration devices, usually mechanical surface aerators, to maintain aerobic conditions in the lagoon. The objective is to treat high-strength wastes while retaining the advantages of oxidation ponds, i.e. rapid B.O.D. removal and no odour. Because of the aeration devices, however, bacterial solids (sludge) remain in suspension (providing rapid B.O.D. removal), but need to be removed from the effluent prior to discharge. This can be achieved using either a settling tank or by passing the effluent from the aerated lagoon into an oxidation pond.

Advantages:	low capital costs; no odour; rapid B.O.D. removal.
Disadvantages:	large area required; solids removal is required prior to discharge; relatively high operating costs.

Activated sludge

This is a high-rate aerobic treatment method which uses a high concentration of bacteria to achieve the high rate. In concept, it is a further development of an aerated lagoon, and is shown diagrammatically in Figure 9.3. The liquid waste is fed into the activated sludge tank where it is vigorously mixed and aerated. Just prior to entry to the tank, the waste is "seeded" with activated sludge, i.e. bacterial cells, which will grow rapidly and degrade the organic material. In the tank, the bacteria remain in suspension, in floc form, and increase in numbers at the expense

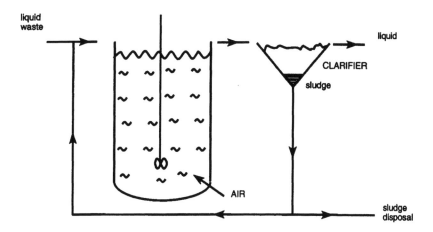

Figure 9.3 An activated sludge unit.

of the B.O.D. On leaving the tank, the effluent enters a clarifier where the sludge rapidly settles out, leaving a clear liquid for disposal. The sludge is drawn off from the base of the clarifier; some is recycled to seed the incoming waste, and so achieve the high bacterial concentration in the tank, while the remainder is sent for disposal.

The activated sludge process can achieve high rates of B.O.D. removal with high-strength wastes, but it can suffer from two major problems. First, if the sludge does not settle properly in the clarifier (a problem known as "bulking") solid material may be discharged with the liquid effluent. Secondly, the process does not respond well to shock loadings, which can lead to poor treatment and bulking. Hence, balancing tanks may be needed, particularly if the factory operates for only five days per week, to ensure a constant feed to the activated sludge tank.

A major disadvantage of the activated sludge process is that large amounts of excess sludge are produced, and this may constitute a disposal problem. One solution is to extend the detention time in the process from the usual 12–16 hours to, say, 24 hours. In this way, some of the sludge will self-degrade and so minimise the excess.

Advantages: occupies small area; high rate of B.O.D.
                      removal.
Disadvantages: high capital costs; high operating costs;
                      excess production of solids; care is needed
                      to avoid fluctuations in waste load.

Biological filtration (trickling filters)

This is another high-rate aerobic treatment method which uses a high concentration of bacteria to achieve the high rate. But in this case, the high concentration is achieved by adsorbing the bacterial cells onto an inert solid support. This forms a microbial (bacterial) slime over which the waste liquid is passed.

In the past, a trickling filter consisted of a circular tank filled with an inert material such as coke or rocks. The microbial slime grew on the inert material, so that when the liquid waste was allowed to filter through, the bacteria grew at the expense of the organic material. Hence the B.O.D. was reduced. However, due to growth of the bacteria, bits of slime continually sloughed off the filter bed and passed through in the effluent. Thus, after filtration, the effluent often needed to be treated in a clarifier to remove the solids before disposal. Figure 9.4 shows a diagrammatic representation of a trickling filter.

In recent years, plastic packing has become more popular at the expense of conventional stone or coke media. Because of the strength and lightness of these packings it is possible to build tall structures with the packing up to 10 metres deep. Otherwise, modern filters operate in the same way as traditional ones.

Advantages: occupies small area; low operating costs;
                      suitable for moderately strong wastes.
Disadvantages: high capital costs; excess production of
                      solids; care is needed to avoid fluctuation
                      in waste load.

Compared to the activated sludge process, biological filtration will give a lower B.O.D. removal, but the operating costs are lower, as are the maintenance requirements. Similar to the activated sludge process, trickling filters do not cope well with fluc-

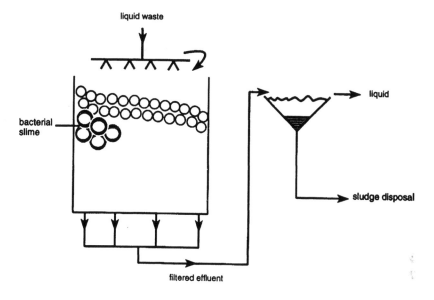

liquid waste

bacterial slime

liquid

sludge disposal

filtered effluent

Figure 9.4   Biological filtration (trickling filter).

tuating waste loads, so a system of flow equalisation may need to be provided prior to treatment of the waste. During their operation it is important that trickling filters be kept wet at all times otherwise the microorganisms may die.

A common problem with trickling filters is that of "ponding", whereby strong wastes will cause excessive bacterial slime growth, causing the filter to block. One solution to this problem is to dilute the incoming waste with some of the treated effluent. Another approach is to use two filters in series. When the first becomes blocked, the order of the filters is reversed. This allows the blockage to self-degrade since the blocked filter now receives a waste stream relatively devoid of nutrients. When blockage of the alternate filter occurs, the order of the filters is again reversed, to restore the original situation.

Upward flow anaerobic sludge blanket (UASB)

The two high-rate processes described above are both aerobic processes. Aerobic processes have the advantage over anaerobic processes of providing faster B.O.D. removal, but they suffer the disadvantages of having higher operating costs (due to the requirement for aeration) and they often produce excess solids which require disposal. In contrast, anaerobic processes produce methane gas, which can be used as a fuel, and this can help to offset the operating costs of these systems. The UASB is a recent development which aims to combine a high-rate anaerobic treatment method with low operating costs, with provision to collect methane and use it as a fuel in the factory. The principle of the UASB is to achieve a high concentration of anaerobic bacteria in the system (Figure 9.5).

The liquid waste stream is fed into the bottom of the tank, from where it moves upwards through a dense layer of bacteria, i.e. the sludge blanket. The organic material is degraded, mainly to

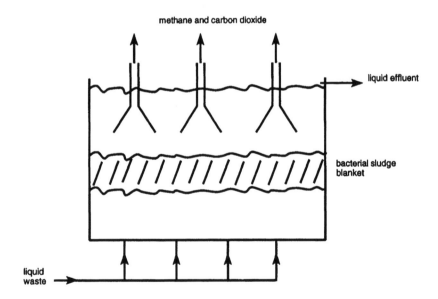

Figure 9.5 An upward flow anaerobic sludge blanket (UASB) unit.

methane and carbon dioxide which are collected at the top of the vessel. The design of the vessel is such that the "blanket" should not flow out of the system with the liquid.

This system has been used successfully for the treatment of abattoir wastes, but considerable expertise is apparently required to control it, i.e. to achieve a consistently high B.O.D. removal without spilling over solids with the liquid effluent. Nevertheless, the UASB has considerable potential for future development and application to the treatment of food industry wastes.

9.3.3.2 Chemical treatment

For waste streams which contain significant quantities of dissolved protein, chemical treatment is a realistic alternative to biological treatment. This is particularly true if the waste stream also contains large amounts of fat, or if the protein and/or fat can be collected and re-processed in some way in the factory. Addition of acid is the most common technique, and the precipitate can then be collected by sedimentation or by dissolved air flotation. Figure 9.6 illustrates a typical example of chemical treatment combined with air flotation. Sulfuric acid is added to

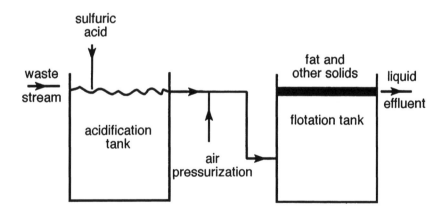

Figure 9.6 Chemical treatment of a waste stream combined with dissolved air flotation.

the waste stream in the acidification tank, causing protein precipitation. The stream is then passed to the flotation tank, being pressurised with air on the way. In the flotation tank, the air bubbles rise to the surface, becoming attached to the precipitate and fat particles as they do so. Finally, the solids are removed from the top of the tank using scrapers, and the liquid effluent can be discharged after pH neutralisation.

Alternative systems use a sedimentation tank, or clarifier, to collect the solids. In this case a flocculating agent may need to be added, along with the acid, to ensure rapid settling of the solids. Polyelectrolytes are commonly used for this purpose.

The main advantages of chemical treatment are the small land area required, and the potential recycling of the solids. Against this, the cost of the chemicals must be considered.

9.3.3.3   Land treatment

In recent years, increasing attention has been paid to this disposal technique. The original stimulus to its use was the concept of putting the liquid effluent on the land, after either primary or secondary treatment, so that the pollutional material could be removed by the plants and soil microorganisms. The liquid would then enter the stream/river system in a much purified form. In addition to this, however, it is now realised that land treatment is also a form of irrigation, particularly in summer months, and a source of fertilizer. The latter, in particular, is receiving increasing attention. Several methods of land treatment are available.

Spray irrigation

This is a useful system for applying liquid effluent to pasture, and involves the use of sprinklers to spray the waste stream onto the land. Normally, the sprinklers are 2–3 m high so that a reasonable "throw" of liquid can be achieved. At the same time, the waste is aerated as it passes through the air. If spray drift is a problem, e.g. close to residential areas, downward directed sprays can be used. The sprinklers are moved at regular intervals to ensure wide coverage and to prevent waterlogging.

Border and furrow method

This system can be used where the soils and topography are suitable. The key to the system is meticulous attention to land shaping to give the uniform low slope required for waste application and the prevention of ponding (Figure 9.7). The waste stream is led into the furrow, which contains a series of gates. These gates can be opened or closed depending on which border it is wished to irrigate. The waste stream flows across the border, allowing water and nutrients to pass into the soil. When one border has been thoroughly irrigated, the waste stream is led to the next one. Ditches generally surround the entire land area to collect the excess liquid and lead it to a receiving water.

The major problem with effluent irrigation systems is that of waterlogging in wet weather. If no other means of effluent disposal is available to the factory, there is a danger of polluting local rivers. Other problems may arise if the waste stream contains pathogenic microorganisms, e.g. in abattoir effluents. In this case, it is important not to run stock on the land, or harvest

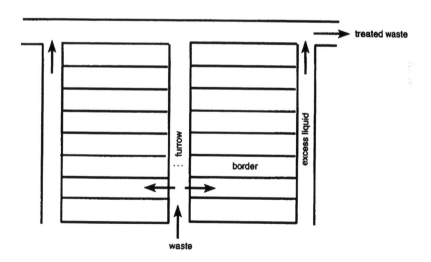

Figure 9.7   Border and furrow irrigation system.

any crops, within a prescribed time period from the last application, e.g. 15 days.

For food factories located in rural areas, it is anticipated that there will be increasing use of land treatment for waste streams. In fact, the waste can be considered as a valuable resource on the basis of its fertiliser value.

9.3.4   Solids Disposal

Waste treatment processes often generate solids, which require disposal, e.g. from primary treatment and activated sludge processes. Disposal of these solids can be expensive and/or difficult depending on the location of the factory. There is no "best" method for treatment, but several options can be considered.

9.3.4.1 Burying

If appropriate land is available, and the transport costs are not too excessive, burial on the land is a reasonable disposal technique.

9.3.4.2 Composting

The objective of composting of solids is to turn a waste material into a valuable fertiliser. The process requires an aerobic treatment of the solids, and so requires aeration, but technology is now available to provide this.

9.3.4.3 Digestion

This is an anaerobic process in which the solids are placed in a digestion tank, and allowed to degrade in the presence of anaerobic bacteria. This reduces the volume of the solids and stabilises them, allowing them to be disposed of on the land or in a lagoon. In addition, a valuable by-product, methane gas, is produced.

9.4   CHOICE OF TREATMENT PROCESS

Except in exceptional circumstances, where there is no legislation governing the discharge of wastes, some kind of waste treatment will have to be conducted. However, the extent of treatment, and the technique used, is usually dependent on individual circumstances.

If a factory's waste stream is to be discharged into a local authority's sewerage system, the authority may place limits on the quantity and quality of the waste discharged. For example, the factory may be compelled to remove all suspended solids from the waste, and to restrict its B.O.D. discharge. At the same time, the factory may pay the authority on the basis of B.O.D. or total solids discharged, so it could be economically advantageous for the company to conduct some kind of treatment of its own.

If the factory discharges its waste stream directly into a receiving water, then the local legislation will determine the extent of treatment required. The type of process used, however, will depend on the individual factory circumstances.

9.5 WASTE MINIMISATION

It is a general objective of all food industries to minimise the effects of wastes on receiving waters, to reduce the amount of waste treatment capacity required, and to reduce the amount of clean water usage and treatment required. This is mainly for economic reasons, but also because of the increasing scarcity of clean water in many countries. Two general approaches can be adopted to achieve these aims, i.e. waste volume reduction and waste strength reduction.

9.5.1 Waste Volume Reduction

If the volume of the waste stream can be reduced, it will be possible to reduce both the capital and operating costs of waste treatment. The following are ways of achieving this:

9.5.1.1 Segregation of wastes

If wastes can be segregated on the basis of their strengths, it may be possible to discharge relatively unpolluted wastes without any treatment whatsoever, thus minimising the volume of waste requiring treatment. The three main classes of wastes are:

- wastes from manufacturing processes. These will generally require some treatment. As a general rule, waste streams from individual processes should be treated separately to ensure their efficient treatment. Only after treatment should individual waste streams be combined;
- waters used as cooling agents. The volumes of these waste streams vary considerably from industry to industry, but they are relatively non-polluted and so require little or no treatment. They should not be used to dilute waste streams from manufacturing processes, as this will increase the overall cost of treatment;
- waters from cleaning procedures. The volumes and strengths depend on the processes and factories concerned, but often these wastes require little treatment.

9.5.1.2 Changing production to decrease wastes

In principle, this is an excellent method to reduce waste volumes, but it is often difficult to put into practice. Examples include:

- improved process control, e.g. turning off water taps when not required
- improved equipment design
- good housekeeping, e.g. regular dry cleaning rather than hosing with water
- preventive maintenance.

9.5.1.3 Conservation and re-use of waste water

If a waste stream is only slightly polluted, e.g. cooling water, it makes little sense to mix it with a high strength waste so that an increased volume requires treatment. It makes more sense to recycle the cooling water so that it can be re-used for the same purpose. Alternatively, waste cooling water may be sent directly to the water treatment plant if its quality is superior to that of the normal raw water supply. The latter applies also to treated wastes. If the standard of waste treatment is high enough, the treated effluent can be recycled rather than discharged into a receiving water.

9.5.1.4 Elimination of slug discharges of process wastes

A slug discharge, or shock loading, is a waste stream of significantly higher volume and strength than normal, and which is discharged to the treatment plant in a short period of time. Several treatment processes cannot cope with these slug discharges, and so there is an adverse effect on both the process and the receiving water. There are at least two methods of reducing the effects of these discharges. The first is for the factory to modify its practices and so minimise the production of slug discharges. The second is for the installation of balancing tanks in which to store the waste streams prior to their being released to the treatment plant continuously and uniformly over an extended period, e.g. 24 hours.

9.5.2 Waste Strength Reduction

This is the second major objective in waste minimisation. If the total amount of pollutional matter can be reduced, considerable savings in the cost of waste treatment will be achieved. The following aspects can be considered.

9.5.2.1 Process changes

All industrial processes should be continually reviewed in order to improve efficiencies and productivities. As part of these reviews, it is essential that the impact of the process on waste generation and treatment be considered. It may, in fact, be more economical to a company to operate a manufacturing process rather less efficiently if this leads to subsequent savings in waste treatment costs. It is for this reason that water usage and waste treatment processes must be considered as integral parts of a factory operation, and the waste treatment facility may, in fact, determine the way in which manufacturing processes are operated.

9.5.2.2   Equalisation of wastes

It was earlier discussed (section 9.5.1.1) how segregation of waste streams can be used as a means of reducing waste volumes. There are, however, circumstances where it may be useful to mix one or more waste streams. This situation arises if these wastes are difficult to treat on their own, but together they are readily degradable. Alternatively, one waste stream may be mixed with another in a balancing tank to prevent slug discharges to the treatment process. It is impossible to generalise about this situation, except to say that it should be studied closely in each individual factory.

9.5.2.3   By-product recovery

The ultimate goal of all factories is to produce no wastes whatsoever. Unfortunately, this goal still remains elusive in many instances. There are, however, examples from the food industry of saleable by-products originating from waste treatment plants. Methane production and use is now standard practice in many places, while the chemical treatment of abattoir wastes to recover protein is becoming increasingly common. In New Zealand, all industrial and potable ethanol is produced by yeast fermentation of de-proteinated whey; this process is economically viable only if considered as a waste treatment operation.

As environmental legislation becomes increasingly strict, and the extent and cost of waste treatment increases, there will be more and more by-products produced from materials currently considered to be wastes. This is the ultimate target of waste treatment.

Chapter 10

OTHER ASPECTS OF SANITATION PROGRAMS

This chapter discusses several aspects of sanitation programs which have not been dealt with elsewhere, but which form an integral part of the overall approach to production of high quality food products. Some of these aspects are self-evident, but, as will be seen in Chapter 11, it is important that they are dealt with in a systematic manner. Hence, the approach to the subject can be as important as the subject itself.

10.1 PEST CONTROL

10.1.1 Rodents

Rodents are pests for three reasons:

- they are a potential source of disease for personnel, a fact which should be impressed on all staff to encourage them to report any sightings
- they are a source of microorganisms to contaminate the food product, particularly if their faeces come into direct contact with product or product-contact surfaces

- they can consume and/or damage raw materials and finished product, thus causing direct economic losses.

There are three major rodents which may be encountered in food factories:

- brown rat (*Rattus norvegicus*). This comes in a variety of colours, but it can be recognised by having 12 mammary glands, a tail shorter than its body length, and by its small ears. It generally lives at floor level, and, because it is a creature of habit, making fixed runs and trials, it is relatively easy to control by baiting;
- black rat (*Rattus rattus*). Again, this comes in a variety of colours, but it can be recognised by having 10 mammary glands, a tail longer than its body length, and by its large ears. It is a good climber and often lives in roofing areas. Because it is not a creature of habit, it can be difficult to control;
- mouse (*Mus musculus*). This rarely travels more than 20 m from its nest, but it is an erratic feeder, and so can be difficult to control.

Rodent control involves three aspects which, if approached in a systematic manner, will reduce the problem of infestation.

10.1.1.1 Elimination of breeding sites

To breed, rodents require sources of food, water and nesting material. Hence, a major objective is to remove all such sources from the company premises. Perhaps surprisingly, it is not in the food processing areas that rodents live and breed, but, rather, in those areas which are considered to be less important from the sanitation viewpoint, e.g. tradepersons' workshops, offices. Hence, these areas must be included in the control program.

Simple precautions to eliminate breeding sites include:

- outdoors — general tidiness, no accumulation of rubbish or unwanted equipment, keep gardens well tended

- indoors — general tidiness, no accumulation of equipment on floor areas, particularly in workshops where rodents can breed unseen, regular emptying of waste bins.

10.1.1.2 Prevention of entry into food processing areas

In practice, very few buildings can be made more than 90% rodent-proof, but there are simple precautions that can be taken to minimise the problem. Potential entrances should be located and blocked, including: doors and windows closed at night; gaps under doors; breaks in screens covering windows and ventilation ducts; grids over floor drains should be in place; trees should not be too close to buildings.

10.1.1.3 Use of poisons

Regular inspections should be made for rodent infestation. Signs include droppings, urine (fluoresces in UV light), gnawed wood and nesting material. Mice can often be detected by their smell. If infestation is found, poisons can be laid, preferably in conjunction with traps. Ultrasound machines are occasionally used to drive off rodents. They can be operated at night, and produce sound of a frequency which rodents can hear, but humans cannot. The idea is that the rodents leave the premises to escape the sound, but these machines may not always be completely effective.

10.1.2   Insects

The major insect pests are flies, cockroaches and ants, but others may occasionally be troublesome depending on the location of the factory. As with rodents, insects are a potential source of disease for personnel, in addition to their ability to contaminate food with their saliva and faeces. Furthermore, the presence of dead insects or their eggs in food or food containers is a reflection of poor standards of sanitation.

Insect control consists of the same approach as with rodents.

10.1.2.1 Elimination of breeding sites

Both indoor and outdoor premises should be kept sufficiently tidy so that insects cannot establish breeding sites. Regular removal of rubbish and good housekeeping procedures will help to solve any problems. Certain factories, e.g. abattoirs, have particular problems because of the stockyards and the abundance of faeces. If the original design of these areas was less than adequate, it is crucial that regular cleaning takes place.

10.1.2.2 Prevention of entry into food processing areas

- All external doors and windows should be kept closed.
- Any open windows should be screened.
- Air curtains can prevent entry of flying insects.
- Bait boxes can be placed in strategic positions outside the buildings so that insects are attracted to these rather than to indoors. These boxes are usually baited with meat, particularly liver, often mixed with insecticide. An example of such a box is shown in Figure 10.1.

10.1.2.3 Killing of insects

- Ultraviolet lights can be used to attract insects to electrified meshes. Insects can see ultraviolet light which humans cannot. When used during the night, red light should be provided for lighting purposes (insects cannot see red light), hence the insects will be attracted to the ultraviolet light and the electrified meshes.
- Insecticides can be used on a daily basis if the insect problem is severe. However, most countries have regulations regarding which insecticides can be used in those areas where food is processed or packaging material is stored. These regulations should always be checked. Normally, the so-called "knock-down" sprays are the only ones approved for food processing areas, and all ingredients, products and packaging materials must be removed from the areas during application.

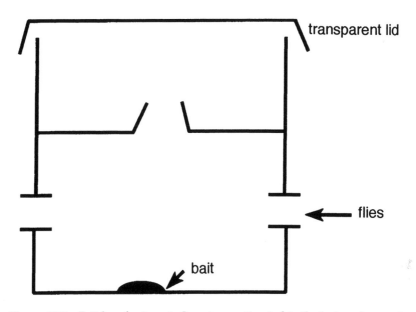

Figure 10.1   Bait box for insects. Insects are attracted to the bait and enter the box through the holes in the sides. To escape from the box, the insects fly upwards towards the light. Because there is no outlet in the lid, the insects become trapped.

Perfumed insecticides or insecticide-impregnated strips are normally not approved for these areas.

When used, insecticides are generally sprayed prior to the final rinse of the cleaning procedure (refer section 7.1). As they are toxic, operators must be properly trained in their use, and provided with protective clothing and goggles. A typical procedure is:

1. Close doors and windows. Switch off ventilation.
2. Spray the quantity of insecticide recommended by the supplier for the volume of the room, ensuring that the spray reaches all parts of the room.

3. Allow the recommended knock-down time (e.g. 10 minutes) before opening the room.

4. Thoroughly rinse all product-contact surfaces.

5. Inspect all equipment to ensure that no dead or dying insects remain.

Other pests which are encountered in the food industry include cats and birds. The control strategies for these are the same as those for rodents and insects.

10.2 PRE-OPERATION HYGIENE INSPECTIONS

In section 7.5, techniques for the evaluation of cleaning procedures were described. As will be discussed in Chapter 11, it is imperative that such evaluation or monitoring techniques are performed systematically and are adequately documented. An example of a way in which this can be done for pre-operation hygiene inspections is now described.

Table 10.1 depicts a Pre-Operation Hygiene Inspection Sheet for a wrapping and packaging area in an abattoir. Every item that is present in the area is included on this checklist. During inspection, each item is viewed, and any contamination defects are noted. These defects may be of three types, based on their likelihood of contaminating the product:

- **critical** (c)   This is a serious contamination defect on a product-contact surface.

- **Major** (M)   This is contamination defect on a product contact surface, but it is an isolated defect; OR, it is a contamination defect in such a position that product is put at risk, e.g. flaking paint on overhead equipment.

- **minor** (m)   This is a contamination defect which is unlikely to put product at risk, e.g. a speck of dirt on a wall.

Table 10.1 Pre-operation hygiene inspection sheet, for a wrapping and packaging area in an abattoir.

Date:				Area:	
Item	m	M	c	Defect	Resolved
Wrap storage cupboard					
Wrapping tables					
Shrink tunnel					
Top belt					
Bottom belt					
Bag loaders					
Vac machine					
Conveyors					
Boneguard table					
Floors					
Walls					
Ceiling					
Handwash					
Soap dispenser					
Towel dispenser					
Item labellers					
QA table					
Carton storage					
TOTAL DEFECTS					
SCORE = 100 – = . SIGNED					

Points are allocated for each type of defect, i.e. critical = 5, Major = 3, minor = 1, and the total defects for the area are determined. This number is subtracted from 100, giving the score for the area. Also included on the inspection sheet are spaces for the defect to be described, and for the action taken to resolve the defect.

Obviously, this inspection sheet has an immediate use in evaluating the effectiveness of cleaning procedures, and ensuring that any remaining dirt is removed before work commences, but it also has a more long-term use. Graphs can be plotted showing the score for each day, and the long-term trends can be monitored. In this way, it can be seen whether the cleaning procedures are generally improving or deteriorating. If the latter, action should be taken to identify the reason, and to remedy the problem.

10.3   PERSONNEL EDUCATION AND COMPANY RULES

Hygienic processing of food and successful implementation of sanitation programs rely on all personnel in the factory knowing exactly what is required of them. There are two approaches to this, i.e the imposition of rules and regulations, detailing the procedures of all work practices, and education of personnel so that they understand the reasons for the rules and regulations. These approaches are complementary, and it is a major aim of management to ensure that both are implemented.

It is not the intention of this book to go into the details of hygiene rules, as these can vary from country to country. However, certain common elements are listed as they form the basis of most companies' rulebooks:

- protective clothing, including hats or hairnets, is compulsory in any area where food is processed or packaging material is stored. This clothing is not to be worn outside of the designated areas;

- footwear must be washed in a sanitiser bath before entering any area where food is processed;

- hands must be kept clean at all times, and must be washed after visits to the toilet;
- cuts and sores must be kept covered with clean dressings at all times;
- smoking is not allowed on the factory premises except in designated areas;
- spitting is forbidden anywhere within the factory premises;
- toilet functions must be performed only in the designated toilet areas.

These basic rules are generally written into a rulebook and are discussed with new staff as part of their induction program. However, in addition to the use of rules, education also plays a part, particularly in curtailing some unhygienic practices against which it is difficult to regulate. For example, it would be desirable to curtail all nose-fingering by staff while they were working in food processing areas, but it would be impractical to establish rules for this. Education programs however, could inform employees of the unhygienic nature of such practices and why they should not be done. Hence, it is the responsibility of management to provide educational programs for all employees, to increase their awareness of basic bacteriology and hygienic practices. The use of videos, cartoons and lectures can all be successful for this purpose, but it is important that all educational programs are planned, and that the company is seen to take the issue seriously.

In the same context, it is the company's responsibility to ensure that all welfare facilities are maintained to the highest possible standard of sanitation, including showers, toilets and dining rooms. While the welfare and comfort of all employees is important, the objective is to achieve the highest possible standard of personal hygiene, and thus prevent contamination of the product.

10.4   PEDESTRIAN TRAFFIC

Control of pedestrian traffic throughout the factory is a major preventative measure that can be taken to minimise contamination of the product. Ideally, factories should be designed and constructed in such a way that traffic control is simplified. In particular, there must be good separation between raw material and finished product, and employee traffic between the different areas must be strictly controlled. In modern factories, different locker rooms and toilet facilities are provided for staff who work in different areas, so that those who work with raw materials do not cross-contaminate those who work with finished products. Also, mechanics and electricians are stationed permanently in one work area, and are not allowed to work in other areas where they may cross-contaminate product. Where it is necessary to allow pedestrian traffic between different areas, strict rules should be laid down governing this movement. Coloured identification tags should be provided on clothing so that supervisors can immediately detect unauthorised personnel in their departments. The rules for the control of pedestrian traffic are straightforward:

- on entering the factory, employees must proceed directly to their locker room
- after changing clothes, employees must proceed directly to their work area
- on leaving the work area, employees must proceed directly to their locker room
- foot baths containing sanitisers must be used where provided.

Visitors to the factory present a special problem. From the sanitation viewpoint, visitors should not be allowed into food processing areas, but from a public relations viewpoint it is often beneficial. All visitors must be provided with the appropriate protective clothing and must be accompanied by company personnel at all times. The route by which visitors are shown around the factory must be carefully planned. As a general rule, always proceed from a "clean" area to a "less clean" area, never *vice versa*. In this way, contamination of product will be minimised.

It must be emphasised that these rules for pedestrian traffic, like all hygiene rules, apply equally to the General Manager as to process workers. If management does not set the example, it is futile to expect the workforce to do so.

10.5 HYGIENIC DESIGN OF EQUIPMENT

Food processing equipment must be designed on the basis of two different requirements:

- the function it is to perform
- the ease of cleaning.

If the equipment is difficult to clean, it may harbour pathogenic and spoilage bacteria, and so affect product safety and quality. Sometimes the requirements of hygienic design may conflict with the functional requirements. In this case, a compromise must be made, but the hygiene requirements must be paramount. The main hygienic design criteria for product-contact surfaces are listed:

1. All materials must be non-toxic, non-absorbent and resistant to product, cleaning chemicals and sanitisers. Stainless steel generally meets these requirements and is widely used.

2. Reinforced plastics and elastomers must not allow penetration of product, and the compression of elastomers must be controlled. Of the plastics, the following are easy to clean, and may be used in hygienic design: polypropylene (PP), polyvinyl chloride unplasticised (PVC), acetal copolymer, polycarbonate (PC), high-density polyethylene (PE).

Polytetrafluoroethylene (PTFE) is a porous material and can often be difficult to clean.

The recommended types of elastomers for use in seals, gaskets and joint rings include: ethylene propylene diene monomer

(EPDM: not fat or oil resistant), nitrile rubber, nitrile butyl rubber (NBB), silicon rubber (suitable for temperatures up to 180°C), fluoroelastomer (suitable for temperatures up to 180°C).

3. The exterior and interior surfaces of all equipment and pipework must be self-draining and easily cleanable. Surfaces should be sufficiently smooth to ensure ease of cleaning.

4. Points to be avoided include: metal to metal joints other than welds, misalignment in equipment and pipe connections, crevices in seals or gaskets, O-rings, screw threads, sharp corners, dead areas.

5. All product-contact surfaces must be easily accessible for visual inspection and manual cleaning, or it must be demonstrated that routine cleaning completely removes all dirt.

One of the most common faults encountered in hygienic design is that it takes too long to dismantle the equipment in preparation for cleaning. When a factory is working to full capacity, and only a limited time is available for cleaning, this can lead to shortcuts being taken with the result that cleaning becomes unsatisfactory. The designers of food processing equipment must be familiar with the hygiene requirements.

10.6   THE ROLE OF THE MICROBIOLOGY LABORATORY

The microbiology laboratory has three roles to play in the food factory:

1. End-product sampling to ensure that it is within specification. In effect, this monitors the processing conditions since the sample that is tested is destroyed during the test. If the sample passes the test, it is assumed that the process is operating effectively and that the untested product is within specification. Conversely, if the sample fails the test, it is assumed that the process is not operating effectively and that the

untested product is also defective. This type of testing is often referred to as accept/reject quality control, and must be conducted when product specifications are imposed by the customer or by government agencies

2. Sampling of end-product, or of intermediate products, to monitor and control the effectiveness of the hygienic procedures operating during processing. This approach is an essential part of the Hazard Analysis and Critical Control Point (HACCP) system described in Chapter 11, and will be discussed further, below. To fully understand the use and value of the approach, however, it should be read in conjunction with Chapter 11

3. Evaluation of cleaning procedures. This has been discussed previously, in Chapter 7.5, and will not be considered further.

10.6.1 Microbiological Assessment of Product During and After Processing

The philosophy here is not to accept or reject product on the basis of the results of the microbiological testing, but to use the results to monitor and control the processing conditions. The concept is sometimes referred to as preventative microbiological quality control, since the aim is to identify and rectify any deficiencies before any real damage is done to the product. In terms of the HACCP system, it is concerned with establishing the critical limits (in this case microbial numbers) to ensure that the process (critical control point) is under control. For the concept to work effectively, the following parameters are required as part of a **sampling plan**.

10.6.1.1 The number of samples tested (n)

The more samples that are taken, the more representative the sampling is, but more expense and labour are involved. A compromise must be made on the basis of the hazard involved, including the future treatment which the product will receive.

10.6.1.2 The microbial limits (m)

The microbiological count is the measure against which it can be judged whether or not the process is under control. If results are obtained which are higher than this value, this is an indication that the process is becoming out of control and that some remedial action should be taken. However, in microbiological counting it sometimes happens that results greater than the limit are obtained, purely because of errors involved in the methodology of counting. In this case, it would not be appropriate to take remedial action if the result were a false alarm. For this reason, a tolerance (c) is sometimes allowed in the microbial limit. While the limit (m) remains the acceptable number of microorganisms, a tolerance is allowed so that a fixed proportion of results are allowed to be above the limit (m), but they must not be greater than the absolute maximum value (M). An example of a sampling plan for testing of beef cuts in an abattoir would be:

$$n = 20$$
$$m = 1 \times 10^5 / cm^2$$
$$c = 4$$
$$M = 1 \times 10^6 / cm^2$$

The acceptable limit of microbial numbers is $1 \times 10^5 / cm^2$. But because of errors involved in the methodology, a tolerance is allowed so that 20% of the samples tested are allowed a count greater than this, provided that they are not greater than $1 \times 10^6 / cm^2$. If greater than 20% of counts are higher than $1 \times 10^5 / cm^2$, or if any count is higher than $1 \times 10^6 / cm^2$, this is a warning that the process is out of control, and remedial action should be taken.

This concept of permitting a tolerance is in agreement with practical experience where, even under good manufacturing practice, a few samples may give above-normal values.

The general procedure for establishing the sampling plan is as follows:

1. Investigational sampling. Examine as many samples as possible from the process in order to determine the normal variations in microbial numbers.

2. Set n and c on the basis of the hazard involved, and m and M on the basis that 80% of samples will be within acceptable limits.

This procedure is rather arbitrary, but it can be modified in the future in the light of experience. An important point, however, is that it allows room for improvement, and this is an important aspect of preventative microbiological quality control, or HACCP.

Sampling plans can be established for a range of different microbial counts:

- aerobic mesophilic counts give some indication of the general microbiological conditions affecting safety and storage life. They are an indicator of the general standard of hygiene and of temperature control during food processing and storage. As a general guideline, the higher the aerobic mesophilic counts, the greater the possibility of pathogens being present, and the shorter the future storage life;

- coliform counts are used as an indication that some untoward event has occurred that may have an adverse effect on the food's safety or storage properties. The presence of coliform organisms indicates that the food has been contaminated by extraneous organic material, e.g. cross-contamination with raw materials, and that pathogens may be present. The presence of *Escherichia coli*, a member of the coliform group, indicates that the food has been contaminated with faecal material, and, thus, pathogens such as salmonellae may be present;

- *Staphylococci* are generally of human origin, and their presence on food is often a reflection of the standard of personal hygiene of food handlers.

In addition to these general microbial counts, which can be used to great value in HACCP systems, the microbiology laboratory has a major role in the detection and enumeration of organisms which can cause a food poisoning syndrome. These organisms are described in Chapter 5, but the methodology of their detection is outside the scope of this book.

Chapter 11

MANAGEMENT OF
SANITATION PROGRAMS

The theoretical concepts of microbial control procedures, and the details of sanitation systems, have been discussed in previous chapters. However, the **implementation**, or practice, of sanitation programs can be a difficult task to accomplish. It is true to say that, in the food industry, most hygiene failures arise from failures in the **management** of sanitation programs. Hence, it is important that management procedures are carefully planned and implemented. This chapter deals with this topic.

11.1   THE HAZARD ANALYSIS CRITICAL CONTROL POINT SYSTEM (HACCP)

The traditional way to determine and improve microbiological quality has been to apply microbiological specifications to ingredients and end-products. Often, these have been accept/reject systems. Because most microbiological tests are destructive in nature, i.e. the sample that is tested is actually destroyed, end-product testing monitors the process rather than the product. Hence, if the sample being tested is within specification, it is assumed that the process is operating correctly, and that the untested product will also be within specification. While this system has its merits, it also has its drawbacks. For example, although it **monitors** the processing conditions, albeit at the end of the production line, it provides no **control** over these conditions. The latter is the basis of the HACCP system. HACCP is a systematic approach to the identification of hazards in a process, and to the control of these hazards. In addition, the system lays down the procedures by which the controls are verified, i.e demonstrated to be actually working, and documented. Thus, the three main ingredients of the HACCP system are:

- **hazard analysis.** This involves identifying all the hazards which can arise in the process, starting with the raw materials, passing through all the processing stages, and including the distribution, retailing and likely use of the end product;
- **determination of critical control points (CCPs).** This involves identifying those points in the production and distribution process at which the hazards can be controlled. If these critical control points are not operating effectively, the hazard may cause the final product to be unacceptable;
- **monitoring of critical control points (CCPs).** This involves devising systems to monitor the CCPs, so that their operation can be verified. Recording of the results of the monitoring procedure is an essential aspect of this stage.

The HACCP system was originally developed to identify microbiological hazards, but it is now used as a general technique to control all food-borne disease, whether it arises from microbiological, chemical or physical hazards. In addition to its role in food safety, the HACCP system can be used to identify and control hazards associated with product spoilage.

The most important aspect of the HACCP system is that it is a **systematic** approach to hazard identification and control and procedure verification. In addition to its role in routine production processes, one of its most valuable attributes is its use in emergency situations. For example, a breakdown of a piece of equipment being used in a production process represents a hazard. The HACCP system will clearly identify and document the procedures which must be followed in the event of such an emergency. In this way, instant "on the spot" decisions do not have to be made, as every contingency has been planned for in advance.

Design and Implementation of an HACCP system involves eight stages.

11.1.1   Construction of the Flow Chart

A detailed flow chart must be prepared for each individual production process. This should be sufficiently detailed so that all potential hazards can be easily identified, e.g.

- sources of all raw materials
- transport methods of raw materials
- storage times and conditions for raw materials
- all process details, including times spent at each stage, temperatures, and all other factors which could affect microbial growth
- alternative procedures in the event of breakdown
- details of all equipment, and bottlenecks in the production line
- details of cleaning procedures.

11.1.2 Hazard Analysis

The flow chart must be carefully examined so that all hazards can be identified, e.g.

- contamination of raw materials with visible dirt or microorganisms
- sources of contamination throughout the entire production and distribution process
- steps during processing, storage and distribution which allow microorganisms to survive or even grow

- processes which lack a procedure that effectively destroys microorganisms.

For each hazard, an estimate should be made of the likelihood of its occurrence, and of its severity.

11.1.3   Critical Control Points (CCPs)

The flow chart is carefully examined to identify those points at which the hazards can be eliminated or minimised. A CCP which will completely eliminate a hazard is known as a CCP1, whereas one which reduces the hazard without eliminating it entirely is known as a CCP2.

Examples of CCPs in a simplified flow chart are shown in Figure 11.1. All cleaning and sanitising procedures are included as critical control points.

11.1.4   Establish Critical Limits for each Critical Control Point

Once the critical control points have been identified, limits must be set for each to ensure that each CCP is under control. For example, for a pasteurisation procedure, the temperature and time must be established which will kill all the unwanted micro-organisms, e.g. hold at 72°C for 30 seconds. For cleaning and sanitising procedures, microbial limits of $< 10^2/cm^2$ are often set. For cooling processes, reducing the temperature from 72°C to 4°C within 2 hours may be the limit. These limits must be set on the basis of the severity of the hazard, the practicalities within the factory, and the future treatment which the product will receive.

11.1.5   Monitoring the Critical Control Points

Having identified the critical control points, and established the limits for each, it is necessary to devise the monitoring procedures to ensure that the limits are achieved. For some, this is relatively easy. For example, temperature sensors and recorders can be used to monitor heat treatment procedures, and chilling and freezing conditions. For other CCPs, however, it may be more

Hazard   CCP

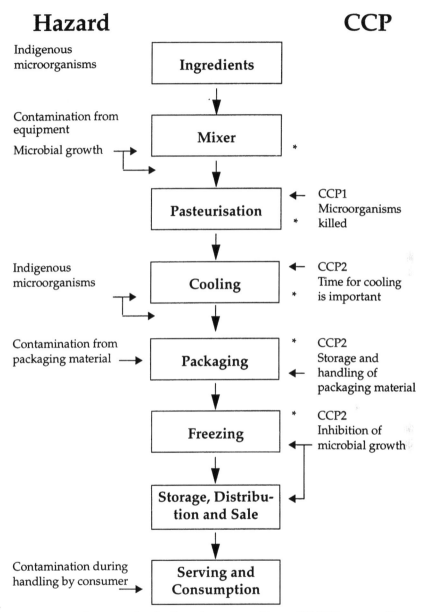

Figure 11.1   An example of a flow chart as used in an HACCP system. (From Harrigan and Park, 1991). Additional CCPs include all points where equipment is cleaned and sanitised (marked *).

difficult, e.g. microbiological testing of product-contact surfaces after cleaning. Here, the results come too late to take any immediate corrective action, but their value is in identifying trends and pinpointing trouble-spots, as described in section 7.5. Microbiological testing of product at various stages of processing should be dealt with in a similar way (refer section 10.6.1).

11.1.6 Establish Procedures for Corrective Action

If the monitoring procedures indicate that a CCP is out of control, it will be necessary to take some corrective action. This action should be pre-planned. For example, if the temperature recorder shows that the temperature of 72°C was held for only 10 seconds instead of 30 seconds, the corrective action for that batch of product should be clearly known. Procedures for corrective action should be prepared for every conceivable malfunction or breakdown in a production process. These procedures must clearly identify those personnel who are responsible for taking the corrective action.

11.1.7 Establish Procedures to Confirm the Effectiveness of the HACCP system

This can include final product testing, but, more importantly, it involves verifying that all of the stages described above are operating in the correct manner. This requires that auditing and checking procedures are developed, and this is a management and supervision function. Traditionally, it is the responsibility of the Quality Assurance Department to monitor quality, but, ideally, all production staff should be involved. Product quality should be seen by all staff as being as important as product quantity. Hence, all monitoring procedures should be clearly understood by all supervisory staff, whose responsibility it is to ensure that the HACCP procedures are being adhered to.

11.1.8   Documentation of Procedures and Records

It is crucial that all monitoring and verification procedures are documented so that there are no misunderstandings or ambiguities regarding these. The documentation should clearly state:

- full details of the monitoring procedures. If the procedure involves laboratory testing, all steps must be clearly described so that the personnel involved know exactly what to do;
- who is responsible for each monitoring procedure;
- the critical limits for each CCP, and under what circumstances action should be taken;
- full details of the procedures for corrective action;
- the documentary procedures to report the action.

If all procedures are correctly documented, the risk of poor decisions being made by inappropriate staff can be substantially reduced.

For most companies, documentation should be produced on four or five levels:

11.1.8.1 The company quality manual

This will include the company's policy, objectives and plan for quality management, and will describe the way in which HACCP is to be carried out.

11.1.8.2 The factory quality manual

This will describe the procedures which are specific to that factory and to the products produced in that factory. All of the descriptions will be sufficiently detailed to avoid misunderstandings. For example, if the procedure is given for the cleaning of a particular department or piece of equipment, sufficient detail must be provided for:

- who will do the cleaning
- when the cleaning will be done

- how the detergent solution should be prepared and at what concentration
- in what order the cleaning should be done, e.g. the top side of a table first, followed by the underside, the legs
- how the rinsing will be done
- how the sanitiser will be prepared and applied
- how the equipment should be left so that it can dry properly
- the monitoring procedure, e.g. how to select which items to assess; how to do the assessment.

11.1.8.3 Departmental procedures

This will be limited to the specific procedures required in each department.

11.1.8.4 Individual instruction sheets

These provide each staff member with a detailed description of all the procedures that he/she is expected to carry out. They should be prepared in collaboration with the individual staff member, and should be written in accurate, non-technical language.

11.1.8.5 Monitoring documents

These include such items as temperature recordings and laboratory notebooks. It is important to accurately record the history of each batch of product so that checks can be made in case of customer complaints. Monitoring documents must be carefully designed so that the appropriate information is recorded, including any actions that were taken.

Finally, a crucial aspect of any HACCP system is that it must be continually reviewed and improved. One of the major problems encountered during the management of sanitation programs is complacency. If a factory is operated for several months without any untoward incidents, there is a danger of believing that it will always be this way. But, beware! The worst accidents happen when management becomes complacent, and the HACCP procedures are not adhered to. Stage 7, described above, which continually monitors that the HACCP system is being correctly implemented, must always be given a high priority.

11.2   THE ISO 9000 SYSTEM

This is a system put forward by the International Standards Organization to provide a framework for a quality management system. **Total quality management** (TQM) is a set of concepts and methods for continuously improving a company and its operations. It ranges from improving the services and materials supplied to the company, through all of the processes within the company, to the degree to which the customers' needs are met. TQM, as a concept, is essentially a set of management techniques which can be used to continuously achieve improvement within the company.

ISO 9000 sets out the requirements of company management which are designed to ensure the quality of products and to provide guidance on topics that must be considered for this purpose. Table 11.1 lists the topics for which procedures must be

Table 11.1   Management system requirements of the ISO 9000 system.

Management responsibility	Inspection, measuring and test equipment
Quality system	Inspection and test status
Contract review	Control of non-conforming product
Design control	Corrective action
Document control	Handling, storage, packaging and delivery
Purchasing	Quality records
Purchaser-supplied product	Internal quality audits
Product identification	Training
Process control	Servicing
Inspection and testing	Statistical techniques

prepared and documented. However, ISO 9000 does not provide specific criteria, only systems. It is the responsibility of the company to define the criteria which they believe will lead to provision of product of desired quality. For a company to receive certification that they have achieved ISO 9000 standard, it must define the criteria by which it intends to operate, document these criteria, and demonstrate to an external assessor that it conforms to these same criteria. Hence, companies define their own standards but they must demonstrate to an external assessor that they conform to them.

Components of the ISO 9000 system include:

- ISO 9000   The guide to selection and use of the standards.
- ISO 9001   This covers aspects of design, development, production, installation and servicing.
- ISO 9002   This covers specifications for production and installation.
- ISO 9003   This covers specifications for final inspection and testing.
- ISO 9004   This is the guide to quality management.

The procedure for a company to gain certification is:

1. Preparation of the documentation regarding the particular ISO 9000 standard.
2. Assessment of the documentation and working practices by an accredited external organisation.
3. Remedial action on any items which show non-compliance.
4. Certification.
5. Maintenance of the system.

It should be noted that the ISO 9000 system does not specifically address the issue of product safety in the way that the HACCP system does. However, it does indicate that a company has a commitment to product quality, and this in itself should instil confidence into customers. If possible, companies should aim to satisfy the requirements of both systems.

REFERENCES AND FURTHER READING

1. EHEDG Update "Hygienic equipment design criteria". *Trends in Food Science and Technology,* July 1993, Volume **4**, 225–229.

2. Harrigan, W.F. and Park, R.W.A. *Making safe food: a management guide for microbiological quality.* Academic Press, London, 1991.

3. Mayes, T. The application of management systems to food safety and quality. *Trends in Food Science and Technology,* July 1993, Volume **4**, 216–219.

4. Sprenger, R.A. *Hygiene for management.* Highfield Publications, Wakefield, England. 1983.

APPENDIX 1

EVALUATION OF DETERGENTS

There are several laboratory tests which can be used as a primary evaluation for the suitability of detergents, but the final decision should be made only on the basis of an actual performance test in which the ability of a product to remove a particular type of dirt from the surface in question is examined. The detergent should not corrode the surface, and must leave no residue after rinsing with water.

The following tests have been devised for simplicity, i.e. to avoid any need for expensive equipment. They apply to alkaline detergents, but it would be possible to devise similar tests for other types of detergents.

1. Ease of Dissolution

The recommended concentration should be dissolved in water, and the ease of dissolution noted using an arbitrary scale.

2. pH

The pH should be determined at the recommended concentration. If the pH of an alkaline detergent is too low, it will not be so effective.

3. Active and Total Alkalinity

(a)   Weigh out 20 g of detergent, and make to 500 ml using deionised water. If the sample contains chlorine, add 5 ml of 0.1 M sodium thiosulphate to neutralise it.

(b)   Titrate a 50 ml sample with 0.5 M HCl using phenolphthalein as indicator. The end-point (pH 8.3) gives the **active alkalinity**, i.e. the amount of alkali present in the sample can be expressed as a percentage of the total sample. Alkalinity is usually given as % Na_2O.

Continue the titration after the addition of methyl orange as indicator. The end-point (pH 4.4) gives the **total alkalinity**. Only the active alkalinity is effective in cleaning.

4. Cloud Point

This is the temperature at which the detergent in solution precipitates. A detergent should not be used at a temperature higher than its cloud point.

5. Emulsifying Properties

Chop 2 g of butter into 15 approximately equal pieces, and place in 5 ml of detergent solution. After shaking for 1 minute, note the degree of emulsification using an arbitrary scale.

6. Corrosion

Prepare small strips of the particular material, e.g. stainless steel, aluminium. After thoroughly cleaning and drying the strips, weigh them and place them in the detergent solution. Hold at an appropriate temperature, e.g. 80°C, for 24 hours, and then remove the strips for examination and reweighing.

7. Sequestering Ability

This measures the ability of the detergent to cope with hard water. Titrate the detergent solution with a 1% (w/v) solution of calcium chloride until the first cloudiness is observed. The more calcium chloride that is required, the greater the ability of the detergent to cope with hard water.

8. Cost

The cost of a detergent is also an important parameter. However, if a lot of manual labour is involved in the cleaning process, the labour cost may greatly exceed the detergent cost. In this case it may be more economical to buy a more expensive detergent if its use reduces the labour cost.

APPENDIX 2

EVALUATION OF SANITISERS

As with detergents, there are several laboratory tests which can be used to evaluate sanitisers, and a simple one is described below. However, such tests measure the effectiveness of a sanitiser only in the laboratory, and not in the factory. Hence, laboratory tests are useful in evaluation only when many products are being considered, and the final assessment should be made only after products have been tested in the factory under practical conditions.

Procedure

1. Prepare a solution of the sanitiser at the recommended concentration.
2. Add a fixed amount of the **test organism**, e.g. *Pseudomonas aerogenes*, to the sanitiser.
3. At appropriate time intervals, e.g. 0, 5, 10, and 20 minutes, withdraw a sample from the sanitiser, and count the number of viable bacteria present using standard microbiological techniques.
4. Assess the sanitiser for its bactericidal affect.

APPENDIX 3

SAMPLING OF POTABLE WATER

Sampling systems should be planned, using a diagram of the reticulation system within the factory. This will ensure that each sample is taken from a different outlet, and that every outlet within the factory is tested on a regular basis. A typical procedure for collecting the sample is:

1. Clean the nozzle of the outlet using a piece of clean tissue.
2. Run the water for a least 1 minute.
3. Collect at least 100 ml of water in a clean, sterile container.
4. Perform the test for the number of coliform organisms using standard microbiological procedures (e.g. using the Membrane Filter method).

Chlorine residuals in potable water should be checked on a daily basis. Standard colorimetric systems are available for this, e.g. Lovibond comparators.

APPENDIX 4

BACTERIOLOGICAL ASSESSMENT OF SURFACES

Traditionally, two distinct techniques are available for this, one a **swabbing** technique, and the other an **impression** technique. The methodology described here is that commonly used in factories which have microbiology laboratories. In addition, however, there is a range of products on the market which come pre-sterilised and pre-packaged, and which can be used without any laboratory facilities. The concepts are the same, however. There are also some techniques available which have been developed to provide results more rapidly than the traditional techniques, but the results may be less quantitative in nature.

Swabs

A metal template is used to ensure that the correct area is swabbed, usually 5 cm^2.

Two sterile swabs are used to swab the area. The first swab is moistened with sterile diluent (e.g. peptone water) and rubbed across the designated area in all directions. The second swab is used dry. (The principle of this technique is that the bacteria on the surface are transferred to the swab. Hence, to ensure reproducible results, it is essential that the swabbing technique is

177

perfected, and performed in a consistent manner.) Both swabs are placed in a 10 ml quantity of diluent, and the handles are broken off in such a manner that the portion contaminated by holding is excluded. (The diluent normally contains several glass beads to assist in dislodging the bacteria from the swabs into the liquid during shaking.) Further dilutions are·prepared and the number of viable bacteria present are counted using standard microbiological techniques.

Impression Technique

This is a technique in which a solid agar surface is held against the surface being examined, and the bacteria are transferred onto the agar surface. After incubation of the agar the bacteria develop into visible colonies. The number of colonies present gives a measure of the number of bacteria which were present on the surface sampled.

However, the technique picks up only a proportion of the bacteria, depending on the type of surface and the manner in which the sample is taken. (This is also true, but to a lesser extent, of the swab technique.) It is not suitable for sampling dirty, wet or rough surfaces. Also, it cannot differentiate between contamination consisting of single cells and that of micro-colonies. Hence, the impression technique is usually considered to be less quantitative than the swab technique.

INDEX

Printed and bound by CPI Group (UK) Ltd, Croydon, CR0 4YY

23/10/2024

01777665-0009